Springer Theses

Recognizing Outstanding Ph.D. Research

Aims and Scope

The series "Springer Theses" brings together a selection of the very best Ph.D. theses from around the world and across the physical sciences. Nominated and endorsed by two recognized specialists, each published volume has been selected for its scientific excellence and the high impact of its contents for the pertinent field of research. For greater accessibility to non-specialists, the published versions include an extended introduction, as well as a foreword by the student's supervisor explaining the special relevance of the work for the field. As a whole, the series will provide a valuable resource both for newcomers to the research fields described, and for other scientists seeking detailed background information on special questions. Finally, it provides an accredited documentation of the valuable contributions made by today's younger generation of scientists.

Theses may be nominated for publication in this series by heads of department at internationally leading universities or institutes and should fulfill all of the following criteria

- They must be written in good English.
- The topic should fall within the confines of Chemistry, Physics, Earth Sciences, Engineering and related interdisciplinary fields such as Materials, Nanoscience, Chemical Engineering, Complex Systems and Biophysics.
- The work reported in the thesis must represent a significant scientific advance.
- If the thesis includes previously published material, permission to reproduce this must be gained from the respective copyright holder (a maximum 30% of the thesis should be a verbatim reproduction from the author's previous publications).
- They must have been examined and passed during the 12 months prior to nomination.
- Each thesis should include a foreword by the supervisor outlining the significance of its content.
- The theses should have a clearly defined structure including an introduction accessible to new PhD students and scientists not expert in the relevant field.

Indexed by zbMATH.

Tushar Kanti Saha

Information-Powered Engines

Doctoral Thesis accepted by Simon Fraser University, Burnaby, Canada

Springer

Tushar Kanti Saha
Department of Physics
Simon Fraser University
Burnaby, BC, Canada

ISSN 2190-5053 ISSN 2190-5061 (electronic)
Springer Theses
ISBN 978-3-031-49120-7 ISBN 978-3-031-49121-4 (eBook)
https://doi.org/10.1007/978-3-031-49121-4

This Springer imprint is published by the registered company Springer Nature Switzerland AG
The registered company address is: Gewerbestrasse 11, 6330 Cham, Switzerland

Supervisor's Foreword

Just over 150 years ago, James Clerk Maxwell formulated a thought experiment that became known as "Maxwell's demon." Imagine a box of gas, divided into two chambers. A "neat fingered" being observes positions of gas molecules and decides whether to open a trap door in the wall separating two chambers. Faster molecules pass to the right, slower ones to the left. Eventually, a temperature difference develops between the chambers, one that can be used to perform work. The second law of thermodynamics is seemingly broken; however, accounting for the work needed to run the measuring device and computations makes the network compatible with the laws of thermodynamics.

In 1929, Szilard introduced a simpler version of Maxwell's thought experiment, with a single particle in a box partitioned into two chambers. Observing which side the particle is on and expanding the volume of that chamber also extracts work. The Szilard engine suggested, for the first time, that information can play a critical role in thermodynamics.

The experiments by Tushar Saha presented in this book concern an information-powered engine, a modern realization of Szilard's thought experiment. Previous experiments focused on showing that the second law of thermodynamics is still obeyed once all costs are accounted for. Tushar's goal was to understand better the engine itself. His version consists of a small bead trapped by a focused, horizontal laser beam. The bead is in water at finite temperature and fluctuates up and down in response to the constant buffeting by water molecules. When the particle fluctuates up, the laser beam moves up, "capturing" the fluctuation. The amount that the beam is lifted is chosen carefully, to make sure that no work is done on the particle by the move. In this way, the bead gradually rises and increases its potential energy. The bead can be released on demand to perform work.

Tushar first studied how to maximize the performance of this new type of engine, showing that the best strategy is to respond to all of the observed up fluctuations. He found that it was possible to extract energy at rates approaching the power output of motors in cells of a similar size. Similarly, the bead could reach speeds comparable to those of similarly sized bacteria. Thus, even though information engines began as thought machines, they perform in ways that begin to compare with biological engines powered by chemical reactions. Tushar next considered the role of noise in measurements. They degrade the performance, of course, because noise might make one react to an "up" fluctuation that is actually "down." But there are subtleties at high levels of noise: the engine can suddenly stop working. However, clever changes to the way observations are used can fix the situation and allow a surprisingly large amount of work to be extracted. Finally, Tushar asked what happens when the environment is out of thermal equilibrium and there are "extra" force fluctuations on the bead. He showed that more work can be extracted—more even than the costs of running the engine. This observation connects to older studies of energy harvesting from noisy environments, giving a new perspective to this field of engineering and paving the way for Maxwell's demon to evolve from a theorist's toy to real and perhaps practical applications.

I urge you to read Tushar Saha's excellent PhD thesis, to find out how this was all done.

Burnaby, BC, Canada
September 2023

John Bechhoefer

Acknowledgments

The last five years have been a roller-coaster ride. Completing the Ph.D. while spending half of the journey during the pandemic would not have been possible without the immense support from my family, friends, and colleagues. Here, I would like to convey my appreciation for their support.

First, I would like to thank my supervisor Dr. John Bechhoefer for providing me with the opportunity to work in his lab. It was a wonderful learning experience to work on the information-engine project. I had the rare opportunity to be involved in the project to understand information engines from the day Dr. Bechhoefer conceived the idea. I never imagined that the pet project for a simple demonstration would one day become my Ph.D. thesis. I have been amazed by Dr. Bechhoefer's ability to build and scale project goals by raising precise questions, among others. I am grateful to him for imparting his wisdom throughout the journey. His words have helped me inside and outside the lab, and I am sure they will also help me with my career ahead.

I thank Dr. David Sivak and Dr. Nancy Forde for being my co-supervisors and providing valuable suggestions for my Ph.D. research. I would especially like to thank Dr. David Sivak for the discussions and for explaining stochastic thermodynamics concepts, among others, to me. I have been very fortunate to be a part of the collaboration under the guidance of Dr. Bechhoefer and Dr. Sivak. Their different perspectives but unified vision provided a great learning environment to gather a broader understanding of my project.

Next, I would like to thank Joseph Lucero and Jannik Ehrich for being on the journey to understand information engines. It has been an exciting experience solving problems with you. I have always enjoyed our discussion, and it's something I will miss. Thank you for making it seamless!

I acknowledge my colleagues in the Bechhoefer lab: Avinash, Jomar, Matthias, Ronja, Lisa, Denesh, Xiao-Yi, Wen-dong, Luis, Prithviraj, Karel, Zak, Sina, David, Amin, Dhruv, and Momcilo. A special thanks to Avinash for helping me get acquainted with Vancouver. I am thankful to David Lee for lending equipment for my experiments and James Lang for IT support. I thank Bryan Gormann, Kenneth Myrtle, and Chang Min Kim for training me to use 3-D printers and equipment in the machine shop. I also acknowledge the support from the members of the physics department: Rose, Kelsey, Ayako, Maegan, Ben, Daria, and Vinisha.

I thank the Department of Physics for the scholarships, and Foundational Questions Institute Fund (FQXi) for supporting me for the past three years, which helped me focus entirely on my Ph.D. project.

I extend my gratitude to Suresh Menon for nurturing my interest in science and for encouraging me to take up research.

I thank Samyuktha for being a constant support and always motivating me, even after being on the other side of the world. Also, thank you for always explaining concepts from biology to me, that made Biophysics a lot of fun.

Lastly, and most importantly, I would like to acknowledge my parents for their love, care, and motivation. None of this would have been possible without their support and belief in me. Maa and Baba, I can never thank you enough!

Parts of This Thesis Have Been Published in the Following Journal Articles

1. T. K. Saha, J. Ehrich, M. Gavrilov, S. Still, D. A. Sivak, and J. Bechhoefer, Information engine in a nonequilibrium bath, Phys. Rev. Lett. **131** (5), 057101 (2023).
2. T. K. Saha, J. N. E. Lucero, J. Ehrich, D. A. Sivak, and J. Bechhoefer, A Bayesian information engine that optimally exploits noisy measurements, Phys. Rev. Lett. **129** (13), 130601 (2022).
3. T. K. Saha, J. N. E. Lucero, J. Ehrich, D. A. Sivak, and J. Bechhoefer, Maximizing power and velocity of an information engine, Proc. Natl. Acad. Sci. **118**, e2023356118 (2021).
4. T. K. Saha, and J. Bechhoefer, Trajectory control using an information engine, Proc. SPIE, **117980L** (2021).

Contents

Chapter 1
Introduction

Engines have shaped the evolution of modern human society. We heavily depend on machines such as mobile phones, cars, and airplanes. Engines are machines that convert some form of energy into useful work. Some of the more common traditional engines include heat engines [1], hydroelectric power plants [2], and wind-powered turbines [3]. Heat engines convert heat, generated from combustion of charcoal, diesel, etc., into mechanical motion [4]. Hydroelectric power plants convert the potential energy of water, stored in a dam, into electricity by driving water turbines [5]. Wind turbines work on a similar principle; they generate electricity from rotating blades, driven by wind energy [6, 7].

Traditional engines were the backbone of the Industrial Revolution [8]. They powered big factories that produced goods in large quantities and heavy vehicles that transported these goods. Most macroscopic engines operate at large energy scales and are usually not affected by the thermal fluctuations in their environment, which are of order $k_B T$ in energy. (Here, k_B is Boltzmann's constant and T is the temperature of the local environment.) By contrast, small molecular motors, which power cellular functions in living organisms, operate against fluctuating forces that are much larger than those produced by these motors. However, they have naturally evolved to work efficiently in such dissipative and noisy environments.

In Fig. 1.1, we show one way to classify the different kinds of engines. Note that the list is not complete, and there are other ways to classify engine types (for example, by the fuel source). But the outline serves to place the main focus of this thesis—energy-harvesting information engines—in the context of engines that have been built and studied.

T. K. Saha, *Information-Powered Engines*, Springer Theses,
https://doi.org/10.1007/978-3-031-49121-4_1

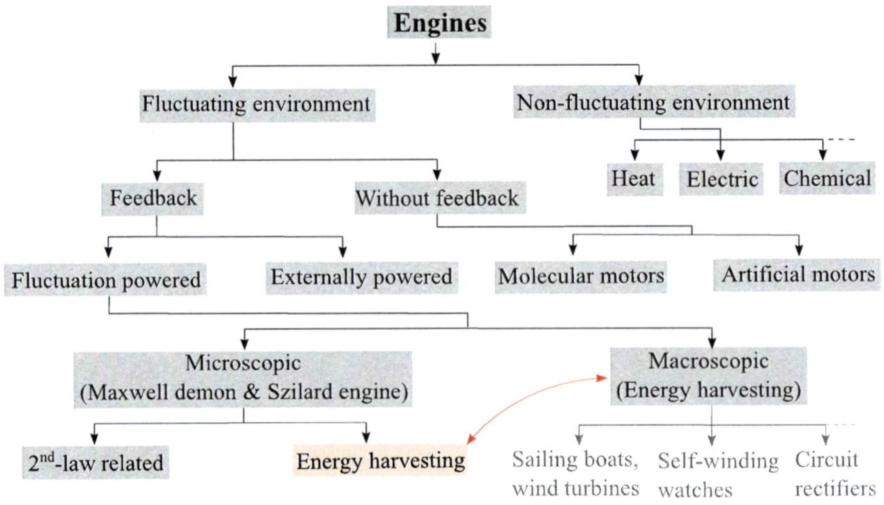

Fig. 1.1 Overview of different kinds of engines. Topic with red shading and arrow denote the focus of this thesis

1.1 Engines in Fluctuating Environments

In this section, we discuss different classes of engines whose operation depend on or are affected by the fluctuations in the environment; see Fig. 1.2. These engines include microscopic engines, such as molecular motors [9, 10] and various types of active matter [11–13], and macroscopic engines, such as sailing boats [14] and wind turbines [7]. These macroscopic engines operate by harnessing energy from the fluctuations in their environment.

1.1.1 Engines Operating Without Feedback

In this section, we discuss engines that operate in a fluctuating environment but do not use feedback. These engines operate by converting some form of chemical energy into motion. Examples range from molecular motors that can generate motion on intracellular scales to ones that are used by multicellular organisms. Molecular motors such as dynein and kinesin can move cargo unidirectionally inside a cell by "walking" along microtubules [15]. They can transport cargo at a speed of $\sim 10^{-6}$ m/s (which, for a 10^{-8} m-long molecular motor, is 10^2 body-lengths per second). A modern car, whose motor is usually about a metre in size, can move at 50 m/s. Although nominally faster, the speed in body lengths per time is comparable to that of a molecular motor [16]. Molecular motors also can be very efficient, even though they operate in a highly viscous surrounding medium. They can operate

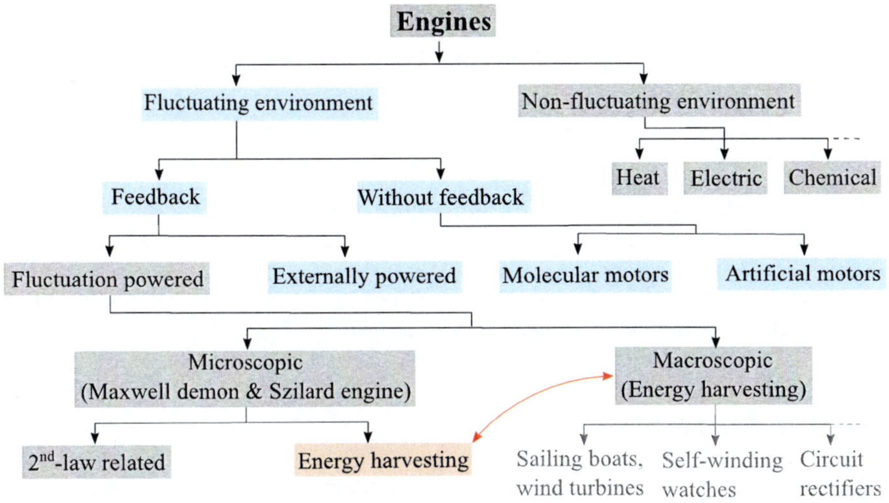

Fig. 1.2 Engines in fluctuating environments. Topics in blue boxes are discussed in this section

at up to 100% efficiency [17–19], whereas cars usually have less than 40% fuel efficiency [20].

Rotary molecular motors in bacteria can generate motion of the single-celled organism. The rotary motor is attached to a flagellum, which rotates to propel the bacterium forward. *E. coli* uses this mechanism to move through a fluid at speeds up to 10 body-lengths per second. Molecular motors can also generate motion in large organisms, by controlling muscle movements. Building blocks of muscles, called sarcomeres, have actin-filament fragments and self-associated myosin-motor filaments [21, 22]. When the organism decides to move, the myosin motors are activated. The head of the myosin motors attaches to the actin filaments and "pulls" the neighbouring acting filaments closer. The regulated motion of the myosin motor leads to muscle movement that assists locomotion in macroscopic organisms.

Molecular motor dysfunction can be fatal for humans: Kinesin motor dysfunction can lead to neurodegenerative diseases [23], and myosin motor dysfunction may lead to heart failure [24, 25]. By understanding molecular motors, we can prevent these diseases [26]. Molecular motors generally run on chemical energy generated by hydrolyzing adenosine triphosphate (ATP) molecules. Other types of molecular motors are fuelled by GTP [27] and proton gradients across membranes [28].

The performance achieved by molecular motors is quite spectacular. The motors continuously collide with molecules in their environment that push them off track. Kinesin motors can achieve unidirectional transport of a cargo by hydrolyzing ATP molecules, using power of the order 10^{-18} W in a highly noisy surrounding medium [29]. Molecular motors have evolved to operate in dissipative environments, with high efficiency. This has motivated research on their working

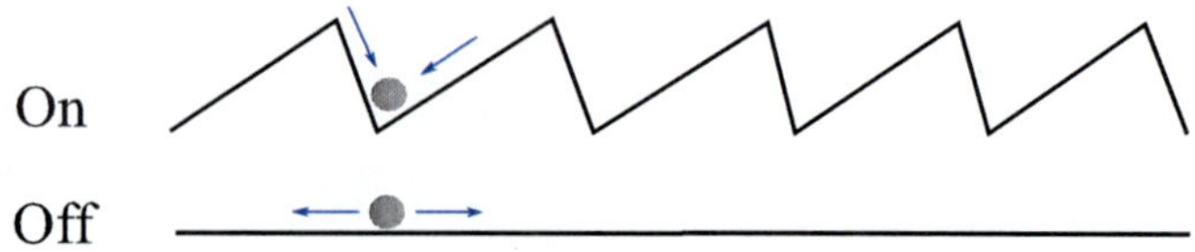

Fig. 1.3 Schematic diagram of diffusing bead in a flashing ratchet. The fluctuations between On and Off states produce an overall motion of the bead from right to left

mechanism, leading to studies that have helped to design artificial microscopic motors and also experimental realizations [30].

Molecular Motor Models

Molecular motors have been modelled as a ratchet [31]. The idea was initially proposed by Lippmann [32] and was further studied by Smoluchowski [33] and Feynman [34]. One model of a ratchet is a periodic asymmetric potential, as shown in Fig. 1.3. The first kind of ratchet is the flashing ratchet [35]. A microscopic system in a static ratchet potential undergoes diffusion and does not move in a particular direction, on average. However, by periodically switching the potential between On and Off states, the bead can be translated in a particular direction. The motion and the direction are controlled by the shape of the ratchet potential. When the potential is turned off, the system undergoes free diffusion. The system dynamics in the Off state of the potential models the state when the head of a kinesin motor unbinds from the actin filament and explores its surrounding. The kinesin-motor head diffuses in the medium until it reaches the next binding site. The kinesin-actin bound state is equivalent to the system in the On-state potential. The system then relaxes to the bottom of the potential. Switching back and forth between the two potentials breaks the detailed-balance criteria and leads to a biased drift-diffusion, driving the system in a particular direction.

Other kinds of ratchet models include tilted and rocking ratchets [36]. In the former, the sawtooth potential is tilted along the direction of transport by adding an external force. In the latter, a rocking force is applied in addition to the sawtooth potential. These forces drive the system out of equilibrium and bias the motion of the system along a particular direction. Some molecular motors are also modelled as having a power stroke [37]. In this case, a static potential is translated in space. The force from the moving potential drags the system along with it.

1.1.2 Engines with Feedback

In the previous section, we discussed ratchet models that can control the dynamics of a system without any feedback. However, the potential landscape can also be switched based on the state of the system in the ratchet potential, as a feedback control. Lopez et al. implemented such a feedback flashing ratchet and showed that it can increase the transport velocity by an order of magnitude over a simple flashing ratchet without feedback [38].

Engines that operate by applying feedback can be broadly described by two categories: externally-powered and fluctuation-powered engines. There are also engines that are powered by a combination of these engine designs. We discuss the fluctuation-powered engines in more detail in Sect. 1.2. Externally-powered engines are designed to apply external forces to a fluctuating system to control its motion. The engine first measures the position of the fluctuating system and then applies feedback forces to direct it towards the desired location. Such systems have been designed using photon nudging[39–41], magnetic steering [42–44], microfluidics [45], etc. These engines control systems in a thermal bath where the fluctuations push the system away from its track; the engine needs to input energy to control the system.

By contrast, there are engines that can convert the thermal fluctuations in the environment directly into useful energy by applying carefully chosen feedback, such that there is zero external work input applied to the motor. This method of using feedback to generate motion (and useful energy), purely from knowledge about the system's fluctuations, is the main theme of this thesis. Such machines form a special class of engines, *information engines*, that are powered purely by thermal fluctuations.

1.2 Fluctuation-Powered Engines

In this section, we discuss engines that operate by converting thermal fluctuations directly into useful energy; see Fig. 1.4. These engines are inspired by an experiment proposed by Maxwell more than 150 years ago to highlight the statistical nature of the second law [46–48]. The thought experiment also leads to an apparent violation of the second law. Thus, we first discuss the experiment proposed by Maxwell and the proposed resolution of the apparent second-law violation. Then we discuss the experiments that realized these engines and discuss why they are consistent with the second law in this context.

1.2.1 Maxwell Demon

James Clerk Maxwell imagined an intelligent being—now known as a "Maxwell demon"—that could watch the motion of the gas molecules in a chamber [46]. The chamber has a partition that the demon can open and close, without friction. As Maxwell had realized in earlier work, molecules of gas at a temperature T do not all move at the same speed, but rather have a distribution of speeds (Maxwell

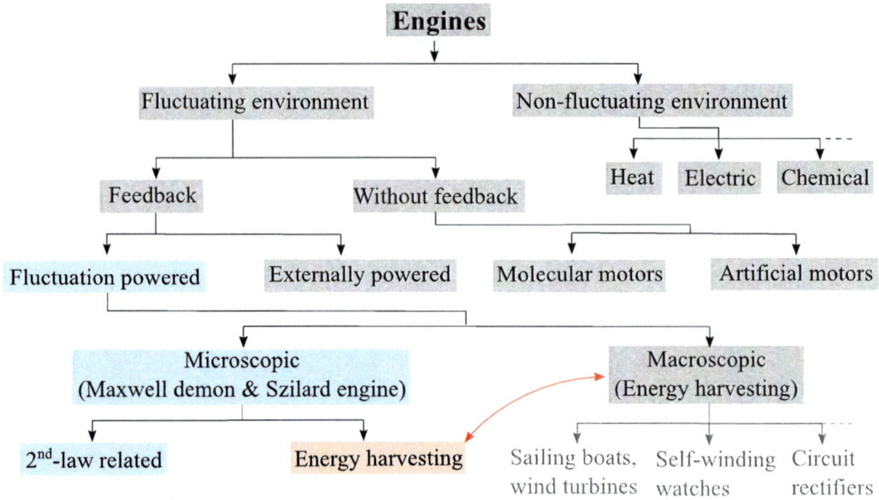

Fig. 1.4 Overview of fluctuation-powered engines. Topics in blue boxes are discussed in this section

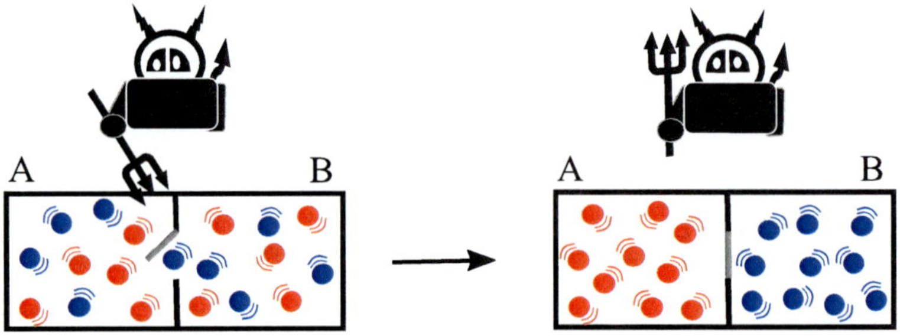

Fig. 1.5 Schematic diagram of a Maxwell demon. The red and blue circles denote the fast and slow moving gas molecules, respectively

distribution) [49]. Thus, some molecules are faster than average, while others are slower than average. The demon opens the gate to allow low-velocity molecules (blue circles) to pass from Chamber A to Chamber B and high-velocity molecules (red circles) in the opposite direction. Sorting all the molecules results in Chamber A's having molecules with higher velocity than Chamber B (Fig. 1.5).

This velocity difference means that the gas in Chamber A has a higher temperature than the gas in Chamber B. The temperature difference between the chambers can be exploited to run a heat engine. This temperature difference, generated by the demon without apparently doing any work, seems to violate the second law of thermodynamics.

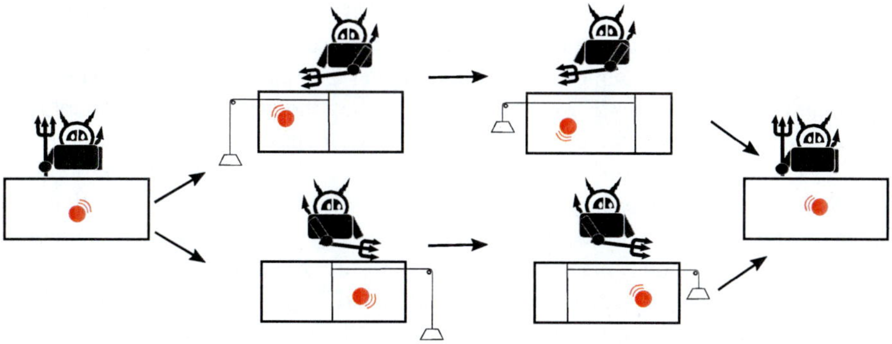

Fig. 1.6 Schematic diagram of a Szilard engine

Szilard Engine

In 1929, Leo Szilard simplified the model of the Maxwell demon [47, 50]. He imagined a single molecule in a gas chamber whose walls are maintained at a constant temperature T. The Maxwell demon inserts a partition into the chamber, and measures whether the molecule is on the left or right side of the partition. The demon attaches a microscopic mass to the side of the partition where the particle is measured and performs a quasistatic expansion. The expansion raises the height of the microscopic mass, hence storing gravitational potential energy. Because the particle interacts with the walls and can transfer energy to and from those walls, it stays in equilibrium with the bath at temperature T (Fig, 1.6).

According to the second law of thermodynamics, the entropy of the universe cannot be reduced. However, Maxwell's thought experiment suggests that it is possible to violate the second law by applying careful feedback based on the velocity of the molecule or, in general, the state of the system. For the Szilard engine, this is equivalent to knowing on which side the molecule resides with respect to the partition. The demon can apparently convert the heat in the medium (the walls) completely into useful energy, either by creating a temperature gradient to run a heat engine or directly converting heat to store gravitational potential energy. The heat extracted in this process then leads to a decrease in the entropy of the universe—thus violating the second law of thermodynamics. This is famously known as the *Maxwell demon paradox*.

Information-Processing Cost

In 1961, Landauer proposed a lower bound on the minimum energy that is required to erase memory (information) from a storage device [51]. The bound depends on the "information content" of the stored memory, which is quantified by the Shannon entropy [52]. In 1982, Bennett proposed that the Landauer erasure cost could resolve the Maxwell demon paradox [53]. To extract energy from a two-state system, such as a Szilard engine where the molecule can be on either the left or right side of the partition, the demon must store one bit of memory. On the one hand, the maximum energy that can be extracted by expanding the gas chamber to double its volume is $k_B T \ln 2$. On the other hand, the Shannon entropy of a one-bit memory is $k_B \ln 2$, and the corresponding Landauer erasure cost is $k_B T \ln 2$. Therefore, the maximum energy that can be extracted from the system is always less than or equal to the information-processing cost, the energy required to erase memory. The energy bounds on work extraction and erasure processes are reached when the transformations are performed in the quasistatic limit. Also, the information-processing cost is the energy required to operate the memory device of the Szilard engine. The energy cost may be associated with either the erasure protocol or the measurement process to correlate the state of the system with the memory [54].

The engines imagined by Maxwell and Szilard were of nanometer scale. Until recently, it was experimentally challenging to build such small devices; however, recent advances in technology and theoretical developments in stochastic thermodynamics [55, 56] have made it possible to experimentally realize information engines. The Landauer cost of erasing a one-bit memory was experimentally verified using a micron-scale bead in a double-well potential [57, 58]. At the start of the erasure protocol, the bead can be in either of the wells, corresponding to the state 0 or 1 of a memory bit, with equal probability. It is assumed that each memory state corresponds to the system's state, being on either the left or right side of the partition, before the work-extraction protocol. The erasure protocol aims to reset the memory (the position of the bead in the wells) to a single state (to one of the wells). A pre-defined protocol manipulates the double-well potential and "pushes" all the probability of the bead's state to a pre-chosen well; e.g., the bead always ends up in the right well after the erasure protocol (with 100% erasure accuracy). In this setting, the erasure cost corresponds to the energy that was required to manipulate the double-well potential that determines the probability distribution. The energy is measured by using stochastic-thermodynamics techniques. We discuss these techniques in detail in Chap. 2. The erasure protocols were performed using finite-time protocols. Extrapolating the results to the infinite-time limit, the experiments [57, 58] verified the thermodynamic energy bound predicted by Landauer. The Landauer bound has been reached in other classical [59–61] and quantum [62–64] systems. Further studies also demonstrated the relation between the Shannon entropy and Landauer cost [65] and found finite-time protocols that reduced dissipation during erasure [66–69].

The experiments that verified the Landauer bound indirectly measured the heat dissipation associated with information processing from changes in the system's potential energy. Koski et al. directly measured the increase in the memory device's local temperature in their electronic Maxwell-demon apparatus. They demonstrated explicitly that information processing is inherently dissipative [70]. These experiments directly resolved the Maxwell demon paradox and inspired further studies on information engines.

1.2.2 Second-Law Related Experiments on Information Engines

In this section, we discuss experimental studies that were carried out to physically realize Maxwell demons and Szilard engines and to clarify the second law of thermodynamics. The first set of experiments demonstrated the existence of the Maxwell demon by realizing such engines in a range of microscopic systems. With theoretical advances in measuring information-processing costs in more complex experimental systems, further studies measured the efficiency of information engines and improved the design to increase efficiency.

Experimental Realizations of Information Engines

Information engines are a modern realization of the Maxwell-demon experiment that rectify fluctuations in a thermal bath. Serreli et al. realized the first molecular information ratchet using a rotaxane (a ring-structured molecule) locked onto a linear polymer [71]. The linear polymer can be in one of two possible configuration, E and Z. In the E configuration of the polymer, rotaxane can diffuse from one end to the other. However, in the Z configuration, rotaxane is locked onto either side of the polymer. The authors can preferentially convert the polymers in the Z configuration into the E configuration when the rotaxane is on the left side of the polymer by shining light on the sample. Then the rotaxane can freely diffuse along the polymer. The polymer is reversed to the Z configuration, and some of the rotaxanes that were initially on the left side end up on the right. Thus, the fraction of rotaxane molecules on the right side of the polymer increases, without changing the chemical potential of the molecule.

A simpler, mesoscale model of the information ratchet was experimentally realized by Lopez et al. [38]. They used an optical line trap to create an array of ratchet potentials. A laser beam was focused inside a sample chamber with micron-sized beads using a microscope objective. The focal spot was oscillated between two points using an acousto-optic modulator. The intensity of the laser was modulated to generate the high- and low-intensity regions that correspond to the potential minima and maxima, respectively. The flashing ratchet was generated by switching the potential between the ratchet and linear potential (without intensity modulation). A feedback algorithm was implemented that switched the potential between the On and Off states, according to the measured position of the bead. The authors showed that the information ratchet could produce average velocities that are an order of magnitude greater than those that can be achieved by a flashing ratchet.

Maxwell-demon-like engines have also been implemented for sorting atoms [72–75]. Thorn et al. [75] implemented the Maxwell demon to separate ^{87}Rb atoms in their ground energy state from those at a higher energy state. The atoms are first cooled to low temperatures (0.1 mK) using a magneto-optic trap and then moved to an optical trap. A pair of laser beams of different wavelengths are aligned perpendicular to the optical trap. The authors choose the wavelength of the first laser to allow atoms in the ground state to pass through and to repel atoms in higher-energy states. The wavelength of the second laser is tuned to excite the atoms to a higher-energy state. Then the pair of lasers passes through the atoms, from right to left. Ones in the ground state do not experience a repulsive force from the lasers; however, the ones in a higher-energy state are repelled and pushed along the direction the laser is moved. This separates the lower-energy atoms from the higher-energy atoms. All the higher-energy atoms are later discarded by flushing the space using the repelling laser. Recent experiments have cooled nanoparticles to millikelvin temperatures using feedback forces that are reminiscent of information engines [76–78].

The experiments discussed in this section demonstrated the existence of information engines in many physical systems; however, they did not quantify the cost of information processing. With advances in the theoretical techniques to quantify information costs in continuous systems [79, 80], later experimental studies focused on building efficient information engines.

Efficient Information Engines

In this section, we discuss some of the experimental studies that measured the efficiency of information engines by calculating the information-processing costs. Toyabe et al. [81] performed the first experimental study to measure the efficiency of an information ratchet. Their information engine increased the free energy of a particle by raising it in a staircase potential, by applying feedback. The experimental apparatus consisted of a dimeric particle of polystyrene beads, fixed at a point using a linker molecule, that undergoes rotational Brownian motion. A periodic potential is applied on the particle, such that the particle experiences an overall spiral-staircase-like potential. Because of thermal fluctuations, the particle transitions between the neighbouring stairs. Without external driving, the particle falls down the stairs, on average, to reduce its potential energy as predicted by the second law of thermodynamics: $\langle \Delta F - W \rangle \leq 0$. Here, ΔF is the change in free energy of the particle, and W is the work done on it.

The authors show that the free energy of the particle can be increased by applying feedback based on the particle's position in the potential. They measured the position of the particle at discrete intervals of time. Every time the particle jumps to a higher-energy state, because of thermal fluctuations, the potential is switched to prevent downward transitions. The work done to switch the potential is kept small compared to the increase in free energy of the particle; thus, $\langle \Delta F - W \rangle > 0$. This is an apparent violation of the second law, but we now understand that one has to account for the information cost, as well.

In 1997, Jarzynski showed that the second law arises by applying Jensen's inequality [82], given by $f(\langle x \rangle) \leq \langle f(x) \rangle$ for a convex function $f(x)$, to the relation $\left\langle \exp\left(\frac{\Delta F - W}{k_B T}\right)\right\rangle = 1$. This equality, now known as the Jarzynski equality [83], relates the equilibrium free energy difference of a system driven arbitrarily quickly from an equilibrium state to some other state, to the work on the system. The Jarzynski equality was verified experimentally by stretching RNA molecules in optical tweezers [84], using a torsional pendulum [85], and in other classical [86–91] and quantum [92] systems. The Jarzynski equality describes the work fluctuations of systems driven out of equilibrium by a pre-defined protocol. However, it does not describe information engines. Toyabe et al. showed that $\left\langle \exp\left(\frac{\Delta F - W}{k_B T}\right)\right\rangle > 1$ for their information engine. Sagawa et al. generalized the Jarzynski equality to incorporate information engines by explicitly accounting for the information cost. The new relation, *generalized Jarzynski equality*, is $\left\langle \exp\left(\frac{\Delta F - W}{k_B T} - I\right)\right\rangle = 1$, where I is the information-processing cost. Toyabe et al. measured the information cost based on Shannon entropy and found that the information-to-energy efficiency of their engine was 28%. They also verified an alternate form of the generalized Jarzynski relation.

Koski et al. verified the generalized Jarzynski equality by measuring the information cost of operating their information engine apparatus [93]. It consists of a single-electron box (SEB) that charges a capacitor connected to the SEB. The latter has two metallic regions, made of different metals, connected by a junction. Electrons move from one region to the other by tunnelling through the junction. The presence of an electron in the second region is detected by a single-electron transistor (SET). Based on the state of the electron in the SEB, the information engine applies a feedback voltage to extract energy from the SEB and charges a capacitor. The authors measure the corresponding information cost from the mutual information between the SEB state and SET measurement. They show that the output of the information engine decreases with measurement noise—we discuss the dependence on measurement noise in more detail later in this thesis. The SEB apparatus could reach a maximum work-to-information efficiency of $\approx 60\%$.

Paneru et al. built an information engine, using a colloidal bead trapped in a harmonic potential, that reached an efficiency of $\approx 98\%$ [94]. The information engine cools the thermal bath by shifting the minimum of the well such that it reduce the potential energy of the bead. Ribezzi et al. built a continuous Maxwell demon using a DNA-molecule attached to two beads in a dual-trap setup [95]. They extracted more than $k_B T \ln 2$ from a two-state system by exploiting rare events. They show that even though the engine can extract more energy, the information-processing cost also increases because they typically need to make many measurements before a rare favourable event is detected. Thus, the combined engine-memory system obeys the second law of thermodynamics.

1.3 Energy-Harvesting Information Engines

Previous experiments focused on building information engines based on Maxwell and Szilard's ideas [38, 70–72, 81, 88, 96–99] to verify the Landauer cost of informa-

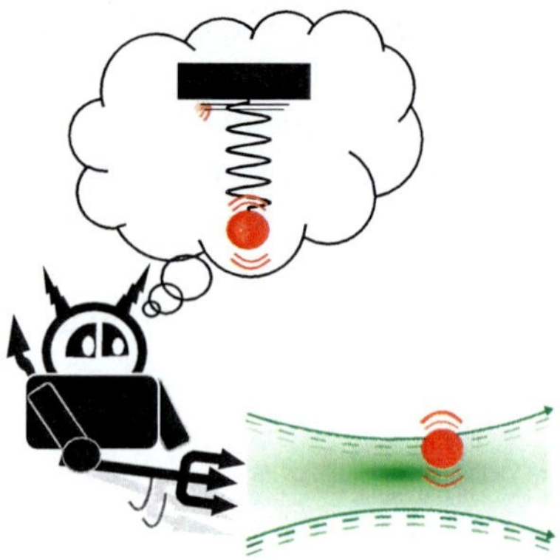

Fig. 1.7 Schematic of our information engine

tion processing [57, 60, 100, 101] and generalized fluctuation theorems [81, 93], and to measure information-to-energy efficiency [93–95, 102–104]. These studies were designed to understand the fundamental physics of operating information engines, in different physical systems. In most cases, the extracted energy could not be used for any practical applications. In this thesis, we build information engines that are designed to harness energy from fluctuating surroundings with maximum performance, in real-world conditions.

We first create and study the performance of an information engine that extracts energy from the thermal bath and *stores* it by raising a weight, as initially imagined by Szilard. The "fuel" for the motor is the information gathered from favourable system fluctuations. Our information engine consists of an optically trapped, micron-scale bead in water (see Fig. 1.7). The laser beam of the trap is horizontal, perpendicular to the vertical gravitational axis. The optical tweezers create a harmonic potential, where the bead fluctuates.

The motion of the heavy bead can be modelled by a simple spring-mass system. The demon monitors the position of the mass and, when the mass fluctuates "up", raises the position of the spring anchor (top bar). Repeating the process raises the mass, by exploiting favourable up fluctuations arising from thermal noise in the medium. The experimental setup is similar to Ref. [94], but here we store the extracted work in a reservoir. The ability to "spend" stored work on demand and for varying purposes increases the utility of the information engine. A previous experimental apparatus introduced by Admon et al. [104] also stored work, but its design was based on a repulsive potential, which meant that the motor was always powered by a combination of external mechanical work and information. Here, with a design based on a trap potential having a local minimum, we ensure that no external work is done on the bead, which simplifies the physical picture. Using this "textbook model" of an information engine, we ask the question, What is the maximum rate of stored gravitational potential energy and directed velocity that the information engine can achieve? We find that the power and velocity of our information engine saturates to a maximum value, even at high measurement accuracy and fast sampling frequency [105]. The ultimate limit is set by the rate of a bead's fluctuation in the trap; above a characteristic value, increasing the sampling frequency of the signal does not significantly improve the engine's performance. The fastest speed achieved by our information engine is $\approx 4 \times 10^2$ body-lengths per second, faster than that achieved by modern cars and comparable to fast marine bacteria [106]. Also, the rate of output energy is comparable to that used by kinesin motors [107].

Although our information engine could generate directed motion, it could not control the bead's trajectory. The random fluctuations that powered the engine lead to a diverging bead position with time. This observation is a consequence of the ratcheting algorithms, which do not control the bead trajectory in space and time. Thus, we introduce an improved feedback algorithm to control the bead's trajectory [108]. Trajectory control makes possible practical applications such as transporting a bead to a specific site at a particular time. Other techniques such as photon nudging [109], fluid flow [45], and magnetic steering [42] can also control the bead trajectory; however, as discussed above, such engines use conventional feedback algorithms that directly apply external forces to the bead. By contrast, information engines are powered by rectifying thermal fluctuations. No additional forces are applied to the bead, and no external work is done: The engine is powered purely by information.

Information engines, in general, are built using sensors with low measurement noise. Previous experiments have shown that inaccurate measurements reduce the performance of information engines because of incorrect feedback [93, 110, 111]. However, measurement noise also reduces the information-processing cost [80] and can increase the efficiency of information engines [110]. With improvements in computational power, especially with FPGAs (field programmable gate arrays) becoming faster and easier to use, one is generally limited by the signal-to-noise ratio (SNR) and not by information-processing capabilities. Thus, we ask the question, For a given noisy detector, how much can we improve the performance of an information engine, if we do not place any limits on the information-processing costs? Indeed, we find that performance can be improved by using better state estimates, calculated from the noisy measurements, for feedback [112].

Information engines operate by rectifying fluctuations available in their environment. In previous experiments, the information engines were designed to rectify the thermal fluctuations associated with the surrounding thermal bath. The finite temperature of the thermal bath sets an upper bound on the performance of the engines: faster measurements and higher information-processing ability did not improve the engine's performance. However, microscopic molecular motors may encounter additional noise because of fluid motion arising from other nearby machines at work in a cell, and these fluctuations can also be rectified by information engines to enhance performance. Thus, we attach a nonequilibrium energy source to the information engine and ask, What are the limits on an information engine's performance that is operating in a nonequilibrium bath? We find that, in such a bath, the performance is not bounded by the temperature of the medium but by the strength and time scale of the energy source.

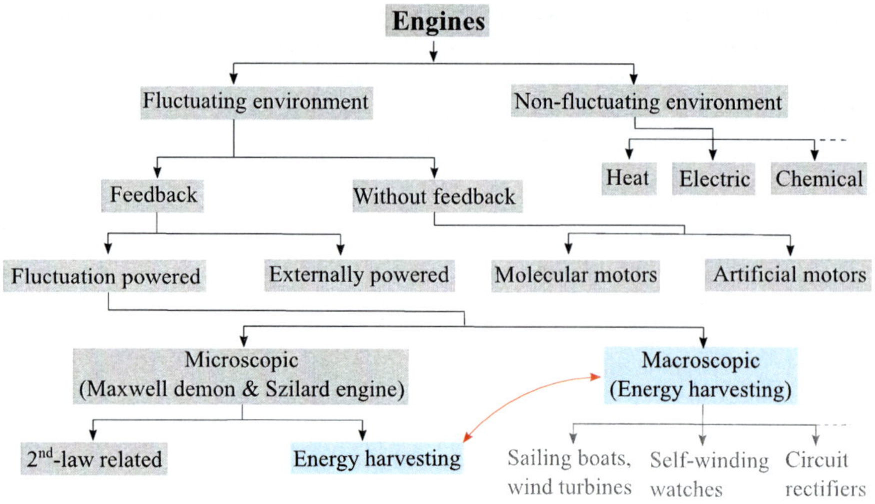

Fig. 1.8 Overview of fluctuation-powered energy-harvesting engines. Topics in blue boxes are discussed in this section

The working principle of information engines is reminiscent of macroscopic engines such as sailboats (Fig. 1.8). They can, in principle, move in any direction by rectifying the direction of the wind flow. They move in an environment that is at a single temperature, unlike heat engines that require two heat baths to operate. Also, there is no fuel cost of running the sailboat because the wind energy is powered "for free" by the Sun; in heat engines such as an internal-combustion engine, one has to burn fuel to generate the temperature difference. The output of a sailboat is limited only by the wind power and the quality of feedback on the sail. Self-winding watches [113] and active electrical current rectifiers [114] are other examples of engines that have autonomous feedback systems that harvest energy, unlike a sailboat, which requires a person to actively steer the direction of the sail based on the direction of the wind. In analogy with the sailboat, one can imagine that information engines that operate in a nonequilibrium environment with sufficiently large fluctuations can in principle extract more energy than they dissipate in information processing.

1.4 Thesis Outline

In this thesis, I present the research work on building information-powered engines that harvest energy that I carried out as a part of my Ph.D. Below is the outline of the topics discussed in this thesis.

In Chap. 2, we discuss the theory concepts that are used in this thesis. I describe the dynamics and energetics of a Brownian particle in an external potential and derive the corresponding discrete-time equations. Then we discuss the statistical distribution of a Brownian particle in a translating external potential and the hydrodynamic flow that the translating particle generates in its surrounding fluid.

In Chap. 3, we discuss the apparatus that I built to perform the experiments presented in this thesis. I first describe the optical feedback tweezers setup and the calibration routines. Then I present the sequential upgrades to the apparatus that were specific to particular experiments that are discussed in Chaps. 4–8.

Chapter 4 introduces the information engine that we built to perform studies that are presented in the rest of the thesis. We first design the feedback algorithm such that the engine solely converts thermal fluctuations into useful energy. Then we study the dependence of the engine's performance on feedback and experimental parameters. Finally, we optimize all the parameters to maximize the performance of the information engine.

In Chap. 5, I introduce a feedback algorithm that can control the position of a bead to track a desired trajectory. The bead is translated without doing any work on it. The thermal fluctuations in the medium are rectified into directed motion by the information engine. I characterize the feedback algorithm and measure its performance limit.

In Chap. 6, I present our study of the performance of the information engine in the presence of measurement noise. We find a reduction in our engine's performance because of measurement noise, as noted previously in other information engines [101, 110]. I implement an inference method, the Kalman filter, that improves the estimate of the bead's position from noisy measurements. Then I use this estimate to apply the feedback introduced in Chap. 4 and show an improvement in the engine's performance.

In Chaps. 4–6, the information engine rectifies the thermal fluctuations from an equilibrium bath, whereas Chap. 7 describes the performance of an information engine in a nonequilibrium bath. We study the dependence of the information engine's performance on the parameters that describe the nonequilibrium noise source.

In Chap. 8, I propose an inference method that can distinguish between information engines and conventionally powered engines. I measure the energy flows, probability distributions, and hydrodynamic flows generated in a medium by bead translation, to contrast information ratchets with power-stroke engines.

Finally, Chap. 9 summarizes the results presented in this thesis and proposes future research on information engines.

References

1. H.W. Dickinson, *A Short History of the Steam Engine* (Cambridge University Press, Cambridge, 2011)
2. P. Breeze, *Hydropower* (Academic Press, Cambridge, 2018)

3. P. Musgrove, et al., Wind power, in *Cambridge Books* (Cambridge University Press, Cambridge, 2009)
4. A. Gupta, S. Narayan, A review of heat engines. Hidraulica **1**, 67 (2016)
5. K.H. Fasol, A short history of hydropower control. IEEE Control Syst. **22**, 68–76 (2002)
6. F. Porté-Agel, M. Bastankhah, S. Shamsoddin, Wind-turbine and wind-farm flows: a review. Bound.-Layer Meteorol. **174**, 1–59 (2020)
7. R.S. Amano, Review of wind turbine research in 21st century. J. Energy Resour. Technol. **139** (2017)
8. N. Rosenberg, M. Trajtenberg, *Studies On Science And The Innovation Process: Selected Works of Nathan Rosenberg* (World Scientific, Singapore, 2010), pp. 97–135
9. A.B. Kolomeisky, M.E. Fisher, Molecular motors: a theorist's perspective. Ann. Rev. Phys. Chem. **58**, 675–695 (2007)
10. R. Iino, K. Kinbara, Z. Bryant, Introduction: molecular motors. Chem. Rev. **120**, 1–4 (2020)
11. M.C. Marchetti, et al., Hydrodynamics of soft active matter. Rev. Mod. Phys. **85**, 1143–1189 (2013)
12. S. Ramaswamy, The mechanics and statistics of active matter. Ann. Rev. Condens. Matter Phys. **1**, 323–345 (2010)
13. M. Bär, R. Großmann, S. Heidenreich, F. Peruani, Self-propelled rods: insights and perspectives for active matter. Ann. Rev. Condens. Matter Phys **11**, 441–466 (2020)
14. Y. An, J. Yu, J. Zhang, Autonomous sailboat design: a review from the performance perspective. Ocean Eng. **238**, 109753 (2021)
15. S.M. Block, Kinesin motor mechanics: binding, stepping, tracking, gating, and limping. Biophys. J. **92**, 2986–2995 (2007)
16. K. Ito, et al., Kinetic mechanism of the fastest motor protein. Chara myosin. J. Biol. Chem. **282**, 19534–19545 (2007)
17. K. Kinosita, R. Yasuda, H. Noji, K. Adachi, A rotary molecular motor that can work at near 100% efficiency. Philos. Trans. R. Soc. Lond. B Biol. Sci. **355**, 473–489 (2000)
18. A.I. Brown, D.A. Sivak, Theory of nonequilibrium free energy transduction by molecular machines. Chem. Rev. **120**, 434–459 (2019)
19. C.-B. Li, S. Toyabe, Efficiencies of molecular motors: a comprehensible overview. Biophys. Rev. **12**, 419–423 (2020)
20. J.A. Caton, The thermodynamic characteristics of high efficiency, internal-combustion engines. Energy Convers. Manag. **58**, 84–93 (2012)
21. J.A. Spudich, The myosin swinging cross-bridge model. Nat. Rev. Mol. **2**, 387–392 (2001)
22. C. Batters, C. Veigel, E. Homsher, J.R. Sellers, To understand muscle you must take it apart. Front. Physiol. **5**, 90 (2014)
23. S.E. Encalada, L.S. Goldstein, Biophysical challenges to axonal transport: motor-cargo deficiencies and neurodegeneration. Ann. Rev. Biophys. **43**, 141–169 (2014)
24. R.J. Hajjar, J.K. Gwathmey, Cross-bridge dynamics in human ventricular myocardium. Regulation of contractility in the failing heart. Circulation **86**, 1819–1826 (1992)
25. D.L. Mann, Y. Urabe, R.L. Kent, S. Vinciguerra, Cooper 4th, G. Cellular versus myocardial basis for the contractile dysfunction of hypertrophied myocardium. Circ. Res. **68**, 402–415 (1991)
26. F.I. Malik, et al., Cardiac myosin activation: a potential therapeutic approach for systolic heart failure. Science **331**, 1439–1443 (2011)
27. E. Schleiff, M. Jelic, J. Soll, A GTP-driven motor moves proteins across the outer envelope of chloroplasts. Proc. Natl. Acad. Sci. U.S.A **100**, 4604–4609 (2003)
28. P.D. Boyer, The ATP synthase—a splendid molecular machine. Ann. Rev. Biochem. **66**, 717–749 (1997)
29. B. Lau, O. Kedem, J. Schwabacher, D. Kwasnieski, E.A. Weiss, An introduction to ratchets in chemistry and biology. Mater. Horiz. **4**, 310–318 (2017)
30. M.R. Wilson, et al., An autonomous chemically fuelled small-molecule motor. Nature **534**, 235–240 (2016)

31. C. Bustamante, D. Keller, G. Oster, The physics of molecular motors. Acc. Chem. Res. **34**, 412–420 (2001)
32. G. Lippmann, The kinetic theory of gases and Carnot's principle. Rep. Congr. Int Phys. Paris **1**, 546–550 (1900)
33. M. Smoluchowski, Experimentell nachweisbare, der üblichen Thermodynamik widersprechende Molekularphänomene. Phys. Z. **13** (1912)
34. R. Feynman, *Feynman and Computation* (CRC Press, Boca Raton, 2018), pp. 63–76
35. F. Jülicher, J. Prost, Cooperative molecular motors. Phys. Rev. Lett. **75**, 2618 (1995)
36. I. Zapata, R. Bartussek, F. Sols, P. Hänggi, Voltage rectification by a SQUID ratchet. Phys. Rev. Lett. **77**, 2292 (1996)
37. W. Hwang, M. Karplus, Structural basis for power stroke vs. Brownian ratchet mechanisms of motor proteins. Proc. Natl. Acad. Sci. U.S.A **116**, 19777–19785 (2019)
38. B.J. Lopez, N.J. Kuwada, E.M. Craig, B.R. Long, H. Linke, Realization of a feedback controlled flashing ratchet. Phys. Rev. Lett. **101**, 220601 (2008)
39. A.P. Bregulla, H. Yang, F. Cichos, Stochastic localization of microswimmers by photon nudging. ACS Nano **8**, 6542–6550 (2014)
40. B. Qian, D. Montiel, A. Bregulla, F. Cichos, H. Yang, Harnessing thermal fluctuations for purposeful activities: the manipulation of single micro-swimmers by adaptive photon nudging. Chem. Sci. **4**, 1420–1429 (2013)
41. N.A. Söker, S. Auschra, V. Holubec, K. Kroy, F. Cichos, How activity landscapes polarize microswimmers without alignment forces. Phys. Rev. Lett. **126**, 228001 (2021)
42. R.W. Carlsen, M.R. Edwards, J. Zhuang, C. Pacoret, M. Sitti, Magnetic steering control of multi-cellular bio-hybrid microswimmers. Lab Chip **14**, 3850–3859 (2014)
43. P.J. Vach, S. Klumpp, D. Faivre, Steering magnetic micropropellers along independent trajectories. J. Phys. D Appl. Phys **49**, 065003 (2015)
44. A. Komaee, B. Shapiro, Steering a ferromagnetic particle by optimal magnetic feedback control. IEEE Trans. Control Syst. Technol. **20**, 1011–1024 (2011)
45. D. Kumar, A. Shenoy, S. Li, C.M. Schroeder, Orientation control and nonlinear trajectory tracking of colloidal particles using microfluidics. Phys. Rev. Fluids **4**, 114203 (2019)
46. C.G. Knott, *Life and Scientific Work of Peter Guthrie Tait*, vol. 213 (Cambridge University Press, London, 1911)
47. H. Leff, A.F. Rex, *Maxwell's Demon 2: Entropy, Classical and Quantum Information, Computing* (CRC Press, Boca Raton, 2002)
48. J.C. Maxwell, *Theory of Heat* (Green and Co., Longmans, 1871)
49. J.C.V. Maxwell, Illustrations of the dynamical theory of gases—Part I. On the motions and collisions of perfectly elastic spheres. Lond. Edinb. Dublin Philos. Mag. J. Sci. **19**, 19–32 (1860)
50. L. Szilard, *Maxwell's Demon*, vol. 2, ed. by H.S. Leff, A.F. Rex (IOP Publishing, Bristol, 2003)
51. R. Landauer, Irreversibility and heat generation in the computing process. IBM J. Res. Dev. **5**, 183–191 (1961)
52. C.E. Shannon, A mathematical theory of communication. Bell Syst. Tech. J. **27**, 379–423 (1948)
53. C.H. Bennett, The thermodynamics of computation—a review. Int. J. Theor. Phys. **21**, 905–940 (1982)
54. I.J. Ford, Maxwell's demon and the management of ignorance in stochastic thermodynamics. Contemp. Phys. **57**, 309–330 (2016)
55. K. Sekimoto, Kinetic characterization of heat bath and the energetics of thermal ratchet models. J. Phys. Soc. Jpn. **66**, 1234–1237 (1997)
56. K. Sekimoto, *Stochastic Energetics* (Springer, Berlin, 2010)
57. A. Bérut, et al., Experimental verification of Landauer's principle linking information and thermodynamics. Nature **483**, 187–189 (2012)
58. Y. Jun, M. Gavrilov, J. Bechhoefer, High-precision test of Landauer's principle in a feedback trap. Phys. Rev. Lett. **113**, 190601 (2014)

59. L. Martini, et al., Experimental and theoretical analysis of Landauer erasure in nanomagnetic switches of different sizes. Nano Energy **19**, 108–116 (2016)
60. J. Hong, B. Lambson, S. Dhuey, J. Bokor, Experimental test of Landauer's principle in single-bit operations on nanomagnetic memory bits. Sci. Adv. **2**, e1501492 (2016)
61. É Roldán, I.A. Martinez, J.M.R. Parrondo, D. Petrov, Universal features in the energetics of symmetry breaking. Nat. Phys. **10**, 457–461 (2014)
62. J.P.S. Peterson, et al., Experimental demonstration of information to energy conversion in a quantum system at the Landauer limit. Proc. R. Soc. A Math. Phys. Eng. Sci. **472**, 20150813 (2016)
63. A.O. Orlov, C.S. Lent, C.C. Thorpe, G.P. Boechler, G.L. Snider, Experimental test of Landauer's Principle at the sub–$k_B T$ level. Jpn. J. Appl. Phys. **51**, 06FE10 (2012)
64. J.V. Koski, V.F. Maisi, J.P. Pekola, D.V. Averin, Experimental realization of a Szilard engine with a single electron. Proc. Natl. Acad. Sci. U.S.A. **111**, 13786–13789 (2014)
65. M. Gavrilov, R. Chétrite, J. Bechhoefer, Direct measurement of weakly nonequilibrium system entropy is consistent with Gibbs–Shannon form. Proc. Natl. Acad. Sci. U.S.A **114**, 11097–11102 (2017)
66. K. Proesmans, J. Ehrich, J. Bechhoefer, Finite-time Landauer principle. Phys. Rev. Lett. **125**, 100602 (2020)
67. P.R. Zulkowski, M.R. DeWeese, Optimal finite-time erasure of a classical bit. Phys. Rev. E **89**, 052140 (2014)
68. G. Diana, G.B. Bagci, M. Esposito, Finite-time erasing of information stored in fermionic bits. Phys. Rev. E **87**, 012111 (2013)
69. J. Melbourne, S. Talukdar, M.V. Salapaka, Realizing information erasure in finite time, in *2018 IEEE Conference on Decision and Control (CDC)* (2018), pp. 4135–4140
70. J.V. Koski, A. Kutvonen, I.M. Khaymovich, T. Ala-Nissila, J.P. Pekola, Onchip Maxwell's demon as an information-powered refrigerator. Phys. Rev. Lett. **115**, 260602 (2015)
71. V. Serreli, C.-F. Lee, E.R. Kay, D.A. Leigh, A molecular information ratchet. Nature **445**, 523–527 (2007)
72. G.N. Price, S.T. Bannerman, K. Viering, E. Narevicius, M.G. Raizen, Singlephoton atomic cooling. Phys. Rev. Lett. **100**, 093004 (2008)
73. A. Kumar, T.-Y. Wu, F. Giraldo, D.S. Weiss, Sorting ultracold atoms in a three-dimensional optical lattice in a realization of Maxwell's demon. Nature **561**, 83–87 (2018)
74. M.G. Raizen, Comprehensive control of atomic motion. Science **324**, 1403–1406 (2009)
75. J.J. Thorn, E.A. Schoene, T. Li, D.A. Steck, Experimental realization of an optical one-way barrier for neutral atoms. Phys. Rev. Lett. **100**, 240407 (2008)
76. T. Li, M.G. Raizen, Millikelvin cooling of an optically trapped microsphere in vacuum. Nat. Phys. 81–110 (2013)
77. J. Gieseler, B. Deutsch, R. Quidant, L. Novotny, Subkelvin parametric feedback cooling of a laser-trapped nanoparticle. Phys. Rev. Lett. **109**, 103603 (2012)
78. F. Tebbenjohanns, M. Frimmer, A. Militaru, V. Jain, L. Novotny, Cold damping of an optically levitated nanoparticle to microkelvin temperatures. Phys. Rev. Lett. **122**, 223601 (2019)
79. T. Sagawa, M. Ueda, Nonequilibrium thermodynamics of feedback control. Phys. Rev. E **85**, 021104 (2012)
80. J.M. Horowitz, H. Sandberg, Second-law-like inequalities with information and their interpretations. New J. Phys. **16**, 125007 (2014)
81. S. Toyabe, T. Sagawa, M. Ueda, E. Muneyuki, M. Sano, Experimental demonstration of information-to-energy conversion and validation of the generalized Jarzynski equality. Nat. Phys. **6**, 988–992 (2010)
82. J.L.W.V. Jensen, On convex functions and inequalities between mean values. Acta Math. **30**, 175–193 (1906)
83. C. Jarzynski, Nonequilibrium equality for free energy differences. Phys. Rev. Lett. **78**, 2690 (1997)
84. J. Liphardt, S. Dumont, S.B. Smith, I. Tinoco, Jr., C. Bustamante, Equilibrium information from nonequilibrium measurements in an experimental test of Jarzynski's equality. Science **296**, 1832–1835 (2002)

85. F. Douarche, S. Ciliberto, A. Petrosyan, I. Rabbiosi, An experimental test of the Jarzynski equality in a mechanical experiment. EPL **70**, 593 (2005)
86. V. Blickle, T. Speck, L. Helden, U. Seifert, C. Bechinger, Thermodynamics of a colloidal particle in a time-dependent nonharmonic potential. Phys. Rev. Lett. **96**, 070603 (2006)
87. S. Schuler, T. Speck, C. Tietz, J. Wrachtrup, U. Seifert, Experimental test of the fluctuation theorem for a driven two-level system with time-dependent rates. Phys. Rev. Lett. **94**, 180602 (2005)
88. G.M. Wang, E.M. Sevick, E. Mittag, D.J. Searles, D.J. Evans, Experimental demonstration of violations of the second law of thermodynamics for small systems and short time scales. Phys. Rev. Lett. **89**, 050601 (2002)
89. O.P. Saira, et al., Test of the Jarzynski and Crooks fluctuation relations in an electronic system. Phys. Rev. Lett. **109**, 180601 (2012)
90. J. Mehl, B. Lander, C. Bechinger, V. Blickle, U. Seifert, Role of hidden slow degrees of freedom in the fluctuation theorem. Phys. Rev. Lett. **108**, 220601 (2012)
91. A. Bérut, A. Petrosyan, S. Ciliberto, Detailed Jarzynski equality applied to a logically irreversible procedure. EPL **103**, 60002 (2013)
92. S. Nakamura, et al., Nonequilibrium fluctuation relations in a quantum coherent conductor. Phys. Rev. Lett. **104**, 080602 (2010)
93. J.V. Koski, V.F. Maisi, T. Sagawa, J.P. Pekola, Experimental observation of the role of mutual information in the nonequilibrium dynamics of a Maxwell demon. Phys. Rev. Lett. **113**, 030601 (2014)
94. G. Paneru, D.Y. Lee, T. Tlusty, H.K. Pak, Lossless Brownian information engine. Phys. Rev. Lett. **120**, 020601 (2018)
95. M. Ribezzi-Crivellari, F. Ritort, Large work extraction and the Landauer limit in a continuous Maxwell demon. Nat. Phys. **15**, 660–664 (2019)
96. P.A. Camati, et al., Experimental rectification of entropy production by Maxwell's demon in a quantum system. Phys. Rev. Lett. **117**, 240502 (2016)
97. N. Cottet, et al., Observing a quantum Maxwell demon at work. Proc. Natl. Acad. Sci. U.S.A **114**, 7561–7564 (2017)
98. Y. Masuyama, et al., Information-to-work conversion by Maxwell's demon in a superconducting circuit quantum electrodynamical system. Nat. Commun. **9**, 1–6 (2018)
99. K. Chida, S. Desai, K. Nishiguchi, A. Fujiwara, Power generator driven by Maxwell's demon. Nat. Commun. **8**, 1–7 (2017)
100. M. Gavrilov, Y. Jun, J. Bechhoefer, Real-time calibration of a feedback trap. Rev. Sci. Instrum. **85**, 095102 (2014)
101. J.V. Koski, V.F. Maisi, J.P. Pekola, D.V. Averin, Experimental realization of a Szilard engine with a single electron. Proc. Natl. Acad. Sci. U.S.A **111**, 13786–13789 (2014)
102. G. Paneru, et al., Optimal tuning of a Brownian information engine operating in a nonequilibrium steady state. Phys. Rev. E **98**, 052119 (2018)
103. G. Paneru, H. Kyu Pak, Colloidal engines for innovative tests of information thermodynamics. Adv. Phys. X **5**, 1823880 (2020)
104. T. Admon, S. Rahav, Y. Roichman, Experimental realization of an information machine with tunable temporal correlations. Phys. Rev. Lett. **121**, 180601 (2018)
105. T.K. Saha, J.N.E. Lucero, J. Ehrich, D.A. Sivak, J. Bechhoefer, Maximizing power and velocity of an information engine. Proc. Natl. Acad. Sci. U.S.A **118**, e2023356118 (2021)
106. G.M. Barbara, J.G. Mitchell, Bacterial tracking of motile algae. FEMS Microbiol. Ecol. **44**, 79–87 (2003)
107. T. Ariga, M. Tomishige, D. Mizuno, Nonequilibrium energetics of molecular motor kinesin. Phys. Rev. Lett. **121**, 218101 (2018)
108. T.K. Saha, J. Bechhoefer, Trajectory control using an information engine, in *Optical Trapping and Optical Micromanipulation XVIII*, vol. 11798 (International Society for Optics and Photonics, Bellingham, 2021), p. 117980L
109. A.P. Bregulla, H. Yang, F. Cichos, Stochastic localization of microswimmers by photon nudging. ACS Nano **8**, 6542–6550 (2014)

110. G. Paneru, S. Dutta, T. Sagawa, T. Tlusty, H.K. Pak, Efficiency fluctuations and noise induced refrigerator-to-heater transition in information engines. Nat. Commun. **11**, 1–8 (2020)
111. A. Taghvaei, O.M. Miangolarra, R. Fu, Y. Chen, T.T. Georgiou, On the relation between information and power in stochastic thermodynamic engines. IEEE Control Syst. Lett. **6**, 434–439 (2022)
112. T.K. Saha, J.N.E. Lucero, J. Ehrich, D.A. Sivak, J. Bechhoefer, Bayesian information engine that optimally exploits noisy measurements. Phys. Rev. Lett. **129**, 130601 (2022)
113. V. Leonov, *Wearable Monitoring Systems* (Springer, Berlin, 2011), pp. 27–49
114. T. Lehmann, Y. Moghe, On-chip active power rectifiers for biomedical applications, in *2005 IEEE International Symposium on Circuits and Systems* (2005), pp. 732–735

Chapter 2
Theory Background

A microscopic particle immersed in a fluid undergoes fluctuating motion in the absence of any visible forces acting on the particle. This phenomenon was first observed by Robert Brown in 1827 while looking at pollen grains in water under a microscope [1, 2]. The jittery motion of microscopic systems is now known as *Brownian motion* [3, Sec. 5.1]. The fluctuating motion of the particle arises from its collisions with the fluid molecules. During such collisions, the Brownian particle exchanges energy with the environment; these energy exchanges are at the heart of our information engine. In this chapter, we first discuss the equation of motion of a trapped Brownian particle, a micron-sized bead. Since we measure the position of the bead at discrete time intervals, we derive the discrete-time equation of motion describing the bead's position from one measurement to the next. The motion of the bead arises from external forces, which could be either deterministic or stochastic. The work done by the external forces is measured by using techniques from stochastic thermodynamics. We then discuss the discrete-time energy estimates that are used in this thesis to measure the energy flows in experiments. Each realization of a stochastic-system evolution follows a different trajectory. These trajectories are best described by a probability density function. Finally, we study the time-dependent evolution of the probability density function of a trapped bead using the Fokker-Planck equation.

2.1 Bead Dynamics

The equation of motion of a Brownian particle—a microscopic bead of radius r and density ρ_{bead}—in an incompressible fluid medium—with density ρ_{water} and dynamic viscosity η—is described by the *Basset-Boussinesq-Oseen equation* (BBO) [4–7],

T. K. Saha, *Information-Powered Engines*, Springer Theses,
https://doi.org/10.1007/978-3-031-49121-4_2

$$m_{\mathrm{f}}\ddot{x}+\gamma\,\dot{x}(t)+\gamma\sqrt{\frac{\tau_{\mathrm{v}}}{\pi}}\int_{t_0}^{t}\frac{\ddot{x}(\tau)}{\sqrt{t-\tau}}\mathrm{d}\tau=\sqrt{2k_{\mathrm{B}}T\gamma}\,\xi(t), \tag{2.1}$$

where $m_{\mathrm{f}} = \frac{4}{3}\pi r^3\left(\rho_{\mathrm{bead}}+\frac{1}{2}\rho_{\mathrm{water}}\right)$ is the effective fluid-bead mass displaced (sum of bead's mass and fluid mass, moving along with the bead where the factor of $1/2$ for the contribution from the surrounding fluid is not important for the purposes of this thesis), $\gamma = 6\pi\eta r$ the bead-water friction coefficient for steady flows, $x(t)$ the bead's position at time t, and $\xi(t)$ a coloured thermal noise that follows $\langle\xi(t)\rangle = 0$, and $\langle\xi(t)\xi(t')\rangle \sim |t-t'|^{-3/2}$ [8]. The "strength" (variance) of the thermal fluctuations depends on the Boltzmann constant k_{B}, temperature T of the thermal bath and γ. The first term on the left-hand side accounts for the additional inertia from the fluid displaced because of the bead's motion and a no-slip boundary condition at the bead-water interface (see Sect. 2.6 for details). The second term is the Stokes drag [9], which accounts for friction on the bead arising from the surrounding fluid. We assume that the bead is in an infinite medium, far enough from any surfaces or other beads that their influence is negligible.

The third term accounts for the hydrodynamic self-interactions of the bead created by unsteady motion in its past in the fluid and is called the Basset history force [10]. As a consequence, the fluctuation in the bead's position is described by a coloured noise [11]. The characteristic timescale of the coloured noise is given by $\tau_{\mathrm{v}} = r^2\rho_{\mathrm{water}}/\eta$. For small bead radii, the correlation time τ_{v} is small and the history force may thus be neglected,[1] and the thermal noise $\xi(t)$ becomes a white noise that follows $\langle\xi(t)\xi(t')\rangle = \delta(t-t')$.

Neglecting the hydrodynamic correlation and adding an external force arising from a potential U, we find the *Langevin equation* [12, 13]

$$m_{\mathrm{f}}\ddot{x}(t) = -\frac{\partial U(x,\lambda)}{\partial x} - \gamma\dot{x}(t) + \sqrt{2k_{\mathrm{B}}T\gamma}\,\xi(t), \tag{2.2}$$

where λ is a parameter that controls the position of the potential U. For micron-sized beads, the inertial relaxation time $\tau_m = m_{\mathrm{f}}/\gamma$ is usually of the order of a microsecond.[2] The experiments presented in this thesis are almost always conducted at a sampling time $t_{\mathrm{s}} \gg \tau_m$, such that the velocity relaxation (inertial effect) term can be neglected, as well. In this regime, relevant for most of the experiments done in this thesis, Eq. (2.2) reduces to the overdamped Langevin equation,

$$\gamma\dot{x}(t) = -\frac{\partial U(x,\lambda)}{\partial x} + \sqrt{2k_{\mathrm{B}}T\gamma}\,\xi(t), \tag{2.3}$$

[1] For a 3-μm silica bead, the vorticity diffusion time is $\tau_{\mathrm{v}} \sim \rho_{\mathrm{water}} r^2/\eta = \frac{10^3\times(1.5\times10^{-6})^2}{8.9\times10^{-4}} = 2.5\,\mu\mathrm{s}$.

[2] The relaxation time for a 3-μm silica bead is $\tau_m = \frac{m_{\mathrm{f}}}{\gamma} = \frac{2/3\,\pi r^3(2\rho_{\mathrm{bead}}+\rho_{\mathrm{water}})}{6\pi\eta r} = \frac{2/3\pi\times(1.5\times10^{-6})^3\times(5\times10^3)}{6\pi\times(8.9\times10^{-4})\times(1.5\times10^{-6})} \approx 1.4\,\mu\mathrm{s}$.

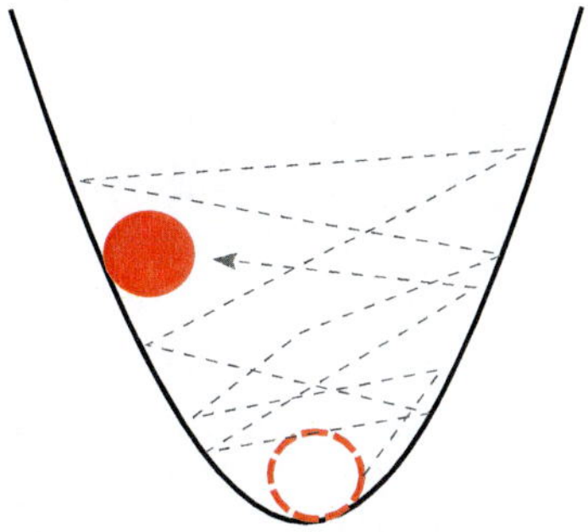

Fig. 2.1 Schematic diagram of a bead, undergoing one-dimentional stochastic dynamics, trapped in a harmonic potential

where $\xi(t)$ is Gaussian white noise that follows $\langle\xi(t)\rangle = 0$ and $\left\langle\xi(t)\xi'(t')\right\rangle = \delta(t - t')$.

Let us consider a simple model where the external force acting on the bead arises from an optical trap (Sect. 2.2). The potential created by the trap can be locally approximated by a harmonic potential $U(x, \lambda) = \frac{1}{2}\kappa\,(x(t) - \lambda(t))^2$ with trap stiffness κ, centred at position $\lambda(t)$ (see Fig. 2.1). The trap applies a linear restoring force on the bead that is given by $F_{\text{trap}} = -\kappa(x(t) - \lambda(t))$. For a horizontally aligned trap (described in Chap. 3), the gravitational force pulls the bead "down". The corresponding overdamped Langevin equation, which accounts for the forces from the trap and gravity, is

$$\gamma\,\dot{x}(t) = \underbrace{-\kappa\,(x(t) - \lambda(t))}_{\text{trap}} - \underbrace{\Delta m\,g}_{\text{gravity}} + \underbrace{\sqrt{2k_{\text{B}}T\gamma}\,\xi(t)}_{\text{thermal noise}}\,, \tag{2.4}$$

where g is the gravitational acceleration. The effective mass $\Delta m = (4/3)\pi r^3\Delta\rho$ of the bead depends on the relative density $\Delta\rho = \rho_{\text{bead}} - \rho_{\text{water}}$ and accounts for buoyancy [14].

2.2 Optical Trap

The optical trap is a technique for holding small particles using optical forces [15, 16]. They are used to trap and manipulate micron-sized beads [17, 18], atoms [19] and single cells [20]. The optical trap presented in this thesis generates a three-dimensional harmonic potential whose centre is moved and controlled along one spatial dimension; see Chap. 3 and Ref. [21] for details of the experimental apparatus.

Optical tweezers are generated by focusing a laser beam using a microscope objective. A microscopic bead with a refractive index higher than the surrounding medium (water) gets trapped near the focus of the laser beam because of optical forces acting on the bead. The model of optical forces depends on the size of the trapped bead (or particle) with respect to the wavelength of the laser. The motion of

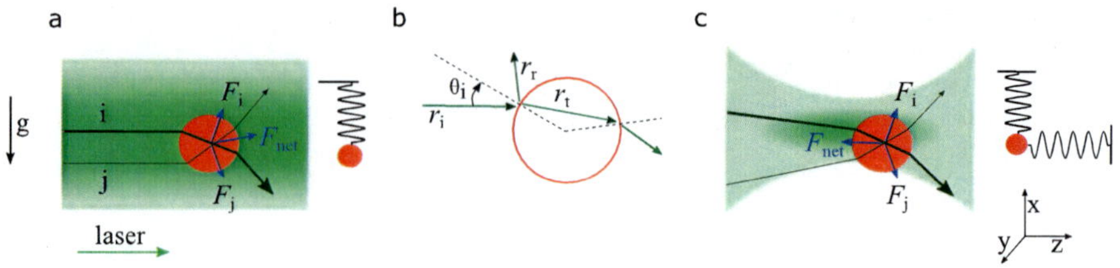

Fig. 2.2 Schematic diagram of a bead trapped in optical tweezers. (**a**) Bead in a Gaussian beam. (**b**) Vectors along the incident ($\mathbf{r}_i$), reflected ($\mathbf{r}_r$), and transmitted ($\mathbf{r}_t$) rays. (**c**) Bead in a three dimensional optical trap generated by focusing the Gaussian beam in (**a**). Blue arrows: direction of force. Green arrows: direction of laser beam rays

trapped particles whose radii are much smaller than the wavelength of the laser can be modelled using the Rayleigh approximation. It treats the particle as a dipole in an electromagnetic field, which experiences a Lorentz force [22]. When the radius of the bead is much larger than the wavelength of the laser, which is the case for most of the experiments presented in this thesis, the trapping can be modelled by simple ray optics [23, 24]. Bead diameters that have size comparable to the wavelength of the laser beam are described by Mie-Lorentz theory [25, 26], which is beyond the scope of this thesis.

In the ray-optics approximation, the laser beam can be described as a collection of many rays, each having a power P_i. The total laser power is $P = \sum_i P_i$, where each ray carries a momentum $n_m P_i / c$ that depends on the refractive index n_m of the medium and velocity of light c. When a light ray strikes a bead at an angle θ_i, most of the light is transmitted along the vector $\mathbf{r}_{i,t}$. A small portion of the light is reflected along $\mathbf{r}_{i,r}$. The transmitted ray is again refracted at the second interface of the bead, as shown in Fig. 2.2a. The force exerted by the ith ray on the bead is given by Ashkin [23]

$$\mathbf{F}_i = \frac{n_m P_i}{c}\hat{\mathbf{r}}_i - \frac{n_m P_i}{c}\hat{\mathbf{r}}_r - \frac{n_m P_i}{c}\hat{\mathbf{r}}_t + \ldots \tag{2.5}$$

where $\hat{\mathbf{r}}_i$, $\hat{\mathbf{r}}_r$, and $\hat{\mathbf{r}}_t$ are the unit vectors along incident, reflected, and transmitted rays whose magnitudes are given by Fresnel's reflection and transmission coefficients, as shown in Fig. 2.2b. For simplicity, we have ignored terms that arising from multiple internal reflections and transmission inside the bead.

The force $\mathbf{F}_i$ arising from the ith ray can be broken down into two components in the plane of incident light. The component along the direction of the laser beam is the scattering force $\mathbf{F}_i^s$, and the perpendicular component is the gradient force $\mathbf{F}_i^g$. To understand the trapping force, consider another ray with force $\mathbf{F}_j$. For a refractive index higher than the surrounding medium, the bead acts as a positive lens. As the light ray i is closer to the centre of the laser, it has a higher power, so that $\mathbf{F}_i > \mathbf{F}_j$. Thus, the bead experiences a net force pulling it towards the centre, to balance the change in momentum. The scattering force is counteracted by the gradient force produced by focusing the laser beam, as shown in Fig. 2.2c. For a heavy bead, the gradient force in x also counteracts the force of gravity. Thus, the bead is trapped after the laser focus, below the centre of the laser beam.

When the size of the bead is smaller than the wavelength of the laser, the optical forces are described by the dipole approximation of the bead. In a homogeneous electromagnetic field E_i, the dipole moment of the bead is given by

$$\mathbf{p} = \alpha_{\mathrm{p}} E(t) , \tag{2.6}$$

where α_{p} is the polarisability of the bead [27]. A dipole in an external electric field experiences a Lorentz force. For a point dipole, one can show that the bead experiences a time-averaged gradient force [26]

$$\mathbf{F}_{\mathrm{grad}}(\mathbf{r}) = \frac{1}{2}\alpha_{\mathrm{p}} \nabla |E_i(\mathbf{r})|^2 . \tag{2.7}$$

The gradient force is again the force responsible for trapping the bead, arising from the potential energy of a dipole in an electric field; hence, it is a conservative force. The particle is also subjected to non-conservative scattering forces [26], which lead to small circulating probability currents [28]. The intensity of the electric field is given by $I = \frac{1}{2} c\, n_{\mathrm{m}} |E|^2$, leading to a gradient force

$$\mathbf{F}_{\mathrm{grad}}(\mathbf{r}) = \frac{1}{4} \frac{\alpha_{\mathrm{p}}}{c\, n_{\mathrm{m}}} \nabla I(\mathbf{r}) . \tag{2.8}$$

For an incident laser with Gaussian laser intensity $I(r') = I_0 \exp\left(-\frac{2r'^2}{w_0^2}\right)$, the gradient force experienced by the bead at small displacements is given by

$$\begin{aligned} \mathbf{F}_{\mathrm{grad}}(\mathbf{r}') &= -\frac{1}{2} \frac{\alpha_{\mathrm{p}}}{c\, n_{\mathrm{m}}} \frac{I_0}{w_0^2} r' \\ &= -\kappa r' , \end{aligned} \tag{2.9}$$

where we have used $I(r') \approx I_0 \left(1 - 2r'^2/w_0^2\right)$, and r' denotes the radial coordinate in the plane perpendicular to the laser propagation. The trap stiffness κ is directly proportional to the laser intensity. Thus, the potential is locally harmonic for small displacements from the trap centre. However, for large displacements, the potential deviates from a harmonic well, as shown in Fig. 2.3.

We use optical tweezers to create a harmonic well around a trapped bead to perform the experiments presented in this thesis. The trap of stiffness κ applying a linear force on the bead is equivalent to a Hookean spring-mass system, as shown in Fig. 2.2. We use the spring-mass system as a (simple) model to understand our information engine.

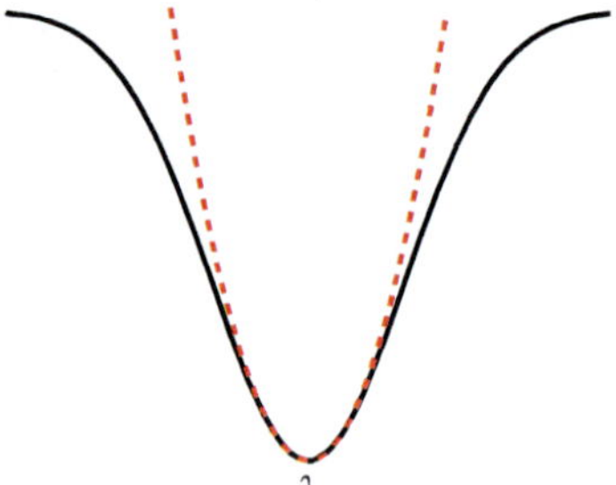

Fig. 2.3 Schematic diagram of harmonic well (red) approximation of the real potential (black) generated by a focused lased beam around a trapped bead

2.3 Discrete Dynamics

In the experiments presented in this thesis, the bead position is measured periodically at discrete time intervals. However, the Langevin equation describes the bead dynamics in continuous time. In this section, we derive the discrete-time dynamics of a bead in a harmonic trap. First, we rewrite the overdamped Langevin equation as

$$\dot{x}(t) = -\frac{\kappa}{\gamma}\left[x(t) - \lambda(t) + \frac{\Delta mg}{\kappa}\right] + \sqrt{2D}\,\xi(t) \tag{2.10}$$

where $D = \sqrt{k_B T/\gamma}$ is the diffusion constant of the bead, and the first term describes the drift of the bead's position because of deterministic forces. Next, we scale the position x and time t as

$$x \to \frac{x}{\sigma} \tag{2.11a}$$

$$t \to \frac{t}{\tau_r}, \tag{2.11b}$$

where $\sigma = \sqrt{k_B T/\kappa}$ and $\tau_r = \gamma/\kappa$ are the natural length and time scales of the bead as it moves in the trap. The scaled equation of motion of the bead is given by

$$\dot{x}(t) = -\left[x(t) - \lambda(t) + \delta_g\right] + \sqrt{2}\,\xi(t)\,, \tag{2.12}$$

where $\delta_g \equiv (\Delta m\ g)/\kappa\sigma$ is the scaled effective mass of the bead. Setting $\dot{x} = 0$ in Eq. (2.12) and averaging over time (with $\langle\xi\rangle = 0$), we see that $\delta_g\,\sigma$ is the length that the equilibrium position of the bead "sags" below the stationary trap centre λ because of gravity. To find the discrete equation of motion, we integrate Eq. (2.12) over a time interval of one sampling time. However, the equation has a stochastic noise term ξ, which cannot be integrated using the standard technique of Riemann sums. Such differential equations are known as stochastic differential equations (SDEs) [3].

We first define $y(t) = x(t) - \lambda + \delta_g$ to simplify the equation to

$$\dot{y} + y = \sqrt{2}\xi(t) \,. \tag{2.13}$$

Note that the trap position λ is assumed fixed during the interval between measurements. (We discuss the feedback timing more precisely in Sect. 3.3.1.) The time integral of the stochastic noise $\xi(t)$ can be formally evaluated as

$$W(t) = \int_0^t \mathrm{d}t' \xi(t') = \int_0^t \mathrm{d}W(t'), \tag{2.14}$$

where $W(t)$ is a *Wiener process* that follows a Gaussian distribution with mean $\langle W(t) \rangle = 0$ and variance $\langle W(t)W(t') \rangle = t - t'$. Substituting the above equation in Eq. (2.13), we find

$$\mathrm{d}y + y\,\mathrm{d}t = \sqrt{2}\,\mathrm{d}W(t). \tag{2.15}$$

Multiplying both sides by $\mathrm{e}^{\int_0^t \mathrm{d}t'} = \mathrm{e}^t$, and integrating the above equation, we find

$$y(t) = y(0)\mathrm{e}^{-t} + \sqrt{2}\mathrm{e}^{-t} \int_0^t \mathrm{e}^{t'} \mathrm{d}W(t'). \tag{2.16}$$

Substituting $y(t) = x(t) - \lambda + \delta_{\mathrm{g}}$ back into the above equation gives

$$x(t) = x_0\mathrm{e}^{-t} + \left(1 - \mathrm{e}^{-t}\right)\left(\lambda - \delta_{\mathrm{g}}\right) + \sqrt{2}\mathrm{e}^{-t} \int_0^t \mathrm{e}^{t'} \mathrm{d}W(t')\,, \tag{2.17}$$

where $x(t = 0) = x_0$. As the bead is driven by Gaussian white noise and linear force in the time interval $[0, t]$, the bead's position is also Gaussian distributed. A Gaussian distribution is completely described by its mean and variance; we thus find the mean and variance of the bead's position distribution to describe its dynamics. Note that integrating the last term in Eq. (2.17) gives a single realization of the bead's position. The distribution of the bead's position over multiple realizations is Gaussian distributed in space at each moment in time.

The mean of the Wiener process follows $\langle dW \rangle = 0$; then, the mean of the integral is

$$\begin{aligned} \left\langle \int_0^t \mathrm{e}^{t'} \mathrm{d}W(t') \right\rangle &= \int_0^t \mathrm{e}^{t'} \left\langle \mathrm{d}W(t') \right\rangle \\ &= 0\,, \end{aligned} \tag{2.18}$$

and the corresponding variance is

$$\left\langle \int_0^t \int_0^t \mathrm{e}^{t'+t''} \mathrm{d}W(t)\mathrm{d}W(t'') \right\rangle = \int_0^t \int_0^{t'} \mathrm{e}^{t'+t''} \left\langle \mathrm{d}W(t')\,\mathrm{d}W(t'') \right\rangle$$

$$= \int_0^t \int_0^{t'} e^{t'+t''} \delta(t' - t'') dt' \, dt''$$

$$= \frac{(e^{2t} - 1)}{2} . \tag{2.19}$$

Defining $x(t) = x(nt_s) \equiv x_n$ and using Eqs. (2.17)–(2.19), we write the discrete-time Langevin equation as

$$x_n = e^{-t_s} x_{n-1} + \left(1 - e^{-t_s}\right) (\lambda_{n-1} - \delta_g) + \sqrt{1 - e^{-2t_s}} \, \xi_{n-1} , \tag{2.20}$$

where ξ_n has zero mean and unit variance. The noise terms at different times are independent: $\langle \xi_m \xi_n \rangle = \delta_{mn}$.

2.3.1 *Itô and Stratonovich Conventions*

In general, we may have to evaluate stochastic integrals of the form

$$I = \int_0^t dW(t) f(x(t)) , \tag{2.21}$$

where $f(x(t))$ is an arbitrary function. The above equation is ill-defined because of the presence of the stochastic noise, which is not a continuous function of time. There are two common conventions used to define the integral: Ito and Stratonovich.

The Itô convention is defined as

$$I_{\text{Ito}} = \lim_{dt \to 0} \sum_{k=0}^{\mathcal{N}} [W(t_k + dt) - W(t_k)] \, f(x(t_k)) , \tag{2.22}$$

where $t_k = k \, dt$ and $\mathcal{N} = t/dt$. In the Ito convention, the function f is evaluated at the beginning of each interval $[t_k, t_k + dt)$.

In the Stratonovich convention, the integral is defined as

$$I_{\text{Str}} = \lim_{dt \to 0} \sum_{k=0}^{\mathcal{N}} [W(t_k + dt) - W(t_k)] \times f\left(\frac{1}{2} [x(t_k + dt) + x(t_k)]\right) , \tag{2.23}$$

with the function f evaluated at the midpoint of the time interval $[t_k, t_k + dt]$. For discrete-time measurements, at time intervals of t_s, the function can be approximated by the average of the function evaluated at the beginning and end of the interval $[t_k, t_k + t_s]$ as

$$I_{\rm Str} = \sum_{k=0}^{\mathcal{N}} [W(t_k + t_s) - W(t_k)] \times \frac{f(x(t_k + t_s)) + f(x(t_k))}{2}. \tag{2.24}$$

For integrals where the noise strength is independent of the position $x(t_k)$ both conventions lead to identical results. However, if the function $f(\cdot)$ depends on position, then the two conventions lead to different results, which might require compensating terms in the stochastic differential equation. In this thesis, we use the Stratonovich convention to calculate quantities such as the heat dissipated by a trapped bead [29].

2.4 Stochastic Thermodynamics

The field of classical thermodynamics describes macroscopic objects in the "thermodynamic limit", where the system's volume and particle numbers tend to infinity. The state of such macroscopic objects is assumed to be described by physical quantities such as temperature, pressure, and volume. These systems are assumed to be in equilibrium: the macroscopic observable quantities such as pressure and density remain constant. Knowing these macroscopic quantities suffices to predict the future states of the system undergoing thermodynamic transformation [29]. These transformations include compression and heat exchange during which the system exchanges energy with the external work source.

The energies exchanged by the system are governed by the laws of thermodynamics. The zeroth law states that when a system is in contact with a heat reservoir, it exchanges energy as heat with the reservoir to reach a new thermal equilibrium. The first law of thermodynamics states that the total energy is conserved. As a consequence, energy flows among the external work source, system, and heat reservoir. The external work source interacts with the system by doing work W to create the thermodynamic transformations. The overall exchange of energy by the system with the external source and the reservoir changes the internal energy U of the system. The classical thermodynamic description of a macroscopic object is deterministic.

The thermodynamic state of the macroscopic system can be understood by studying the dynamics of the constituent microscopic particles. One can use statistical mechanics to show how thermodynamics can arise in systems with a large number of degrees of freedom. At a microscopic level, one must carefully define what is meant by a notion of "hotness". Although the velocity of microscopic particles is correlated with a degree of hotness of the macroscopic object, the state of the microscopic constituents changes rapidly; thus, the state of the microscopic particles is described by a statistical distribution.

A trapped bead in a thermal bath is in a regime that is in between the macroscopic and microscopic descriptions [12]. The state of the bead (system) can be described by coarse-graining the microscopic states arising from the interaction with the heat reservoir. The coarse-grained state is termed a *mesostate*. The interaction of the mesoscopic system with the reservoir leads to randomness in the system state. When a protocol is repeated on a system, the measured trajectory of its state is different each time. The evolution of the system's state $x(t)$ can be modelled either as a time-dependent probability density function or as individual trajectories of a stochastic process. The choice of description depends on the questions of interest being asked of the stochastic system. In general, experiments record individual stochastic trajectories and are best modelled by Langevin equations. The overall characteristics of measured trajectories, however, are best modelled using a probability density function. In Sect. 2.3, we introduced the Langevin equation describing individual trajectories of a trapped bead. In Sect. 2.5, we discuss the time-dependent evolution of the probability density function.

2.4.1 Continuous-Time Energy Estimate

In this section, we discuss the concepts of stochastic thermodynamics that describe the thermodynamics of a single macroscopic-scale object. For simplicity, we start by comparing the components of the stochastic and classical thermodynamics systems. An optically trapped bead in water can be mapped to classical thermodynamic components as follows:

- The bead is the system whose internal energy changes through its interaction with the external work source and heat reservoir. The internal energy is the sum of the kinetic and potential energies of the bead. As the overdamped Langevin equation used to described the trapped bead neglects inertial effects, the internal energy of the bead is due to potential energy only.
- The trap is the external work source. The trap does work on the system by applying a force on a bead that results in its displacement. The time scale of the trap dynamics, i.e., the time taken to update the trap position, is faster than the time scale of the bead dynamics, τ_r.
- Water acts as the heat reservoir. The bead dissipates heat to the reservoir because of viscous dissipation. The bead gains energy from the reservoir when a thermal fluctuation raises its potential energy. The thermal collisions occur at a time scale of picoseconds [30], which is six orders of magnitude faster than the timescale of bead dynamics (microseconds).

The mesoscopic bead undergoes Brownian motion due to thermal fluctuations, as described by the Langevin equation. The bead fluctuates because of the random collisions of water molecules on the bead. Although the Langevin equation does not explicitly account for the collision force from individual molecules, it accounts for the total force from all the colliding molecules, as a stochastic force $\xi(t)$, at every instant in time. This implies that if all the stochastic and deterministic forces acting on the bead are accounted for by the Langevin equation, then it conserves the energy flowing through the bead [31].

Let us start by writing the Langevin equation of the trapped bead as

$$0 = -\gamma \dot{x}(t) + \xi(t) - \frac{\mathrm{d}U}{\mathrm{d}x}\,, \tag{2.25}$$

where the first two terms result from the interaction of the bead with the thermal environment. The last term represents the force acting on the bead because of the trapping potential. These forces sum to zero, implying a force balance in the system. Let the bead be displaced by $\mathrm{d}x$ as a result of thermal fluctuations. Multiplying Eq. (2.25) by $-\mathrm{d}x$ and rearranging the terms we find

$$0 = -\left(-\gamma \dot{x}(t) + \xi(t)\right)\mathrm{d}x + \frac{\mathrm{d}U}{\mathrm{d}x}\mathrm{d}x\,. \tag{2.26}$$

The term $-\gamma \dot{x}(t) + \xi(t)$ quantifies the total energy exchanged between the bead and the thermal bath, which we define as heat $đQ$. The above equation can then be written as

$$đQ = -\left[-\gamma \dot{x}(t) + \xi(t)\right]\mathrm{d}x = -\frac{\mathrm{d}U}{\mathrm{d}x}\mathrm{d}x\,, \tag{2.27}$$

where we write $đQ$ with a slash because its value depends on the particular path $x(t)$ taken by the system. The term $đ$ is used to denote the changes in path-dependent variables, whereas d is used to denote changes in state-dependent variables. Here, the term $đQ$ quantifies a small amount of heat changed arising from the bead displacement $\mathrm{d}x$. Note that the convention used here is that heat dissipated by the bead to the environment is positive. From Eq. (2.27) and conservation of energy overall, we have

$$0 = đQ + \mathrm{d}U\,. \tag{2.28}$$

Equation (2.28) quantifies the energy conservation in a trapped bead system. The heat dissipated by the bead can be estimated without knowing $\xi(t)$, using Eq. (2.27). Heat equals the change in the potential energy of the bead due to the displacement $\mathrm{d}x$, while the external work source U is fixed, as shown in Fig. 2.4. According to

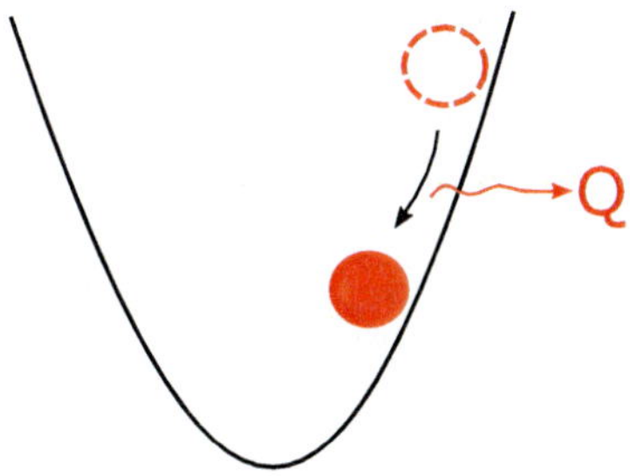

Fig. 2.4 Schematic diagram of energy dissipated as heat Q by the bead to the environment when it moves to a position that has a lower potential energy

Eq. (2.28), energy is interconverted between heat and potential energy of the bead in the trap. The bead absorbs heat, converting it into potential energy when the bead fluctuates up in the potential. The bead dissipates the energy back into the bath as heat when it relaxes back in the potential.

Let us now consider the simple example where the potential U can be controlled by shifting the position λ of the potential-well bottom. Note that the parameter λ can, in general, control any characteristic of the energy landscape, such as the trap stiffness, shape of the potential, etc. The potential energy U is then a function of both the bead's position x and the control parameter $\lambda(t)$. Equation (2.26) becomes

$$0 = -\left[-\gamma \dot{x}(t) + \xi(t)\right] \mathrm{d}x + \frac{\partial U(x,\lambda)}{\partial x} \mathrm{d}x \,, \tag{2.29}$$

where the derivative of the potential is now a partial derivative, with λ held fixed. To make it a total derivative, we add $\frac{\partial U(x,\lambda)}{\partial \lambda} \mathrm{d}\lambda$ to both sides and find

$$\frac{\partial U(x,\lambda)}{\partial \lambda} \mathrm{d}\lambda = đQ + \mathrm{d}U \,. \tag{2.30}$$

The quantity on the left-hand side of Eq. (2.30) is defined as the work $đW$ done by the external work source, the trap in this case, leading to

$$đW = \frac{\partial U(x,\lambda)}{\partial \lambda} \mathrm{d}\lambda \,, \tag{2.31}$$

where $đW$ is again a path-dependent quantity.

The work done by the external source is defined as the change in the potential energy of the bead in the external work source, when the control parameter λ is modified. It is positive when it increases the system's energy, as shown in Fig. 2.5. Substituting the above equation in Eq. (2.30), we reproduce the first law of thermodynamics from the Langevin dynamics,

$$đW = đQ + \mathrm{d}U \,. \tag{2.32}$$

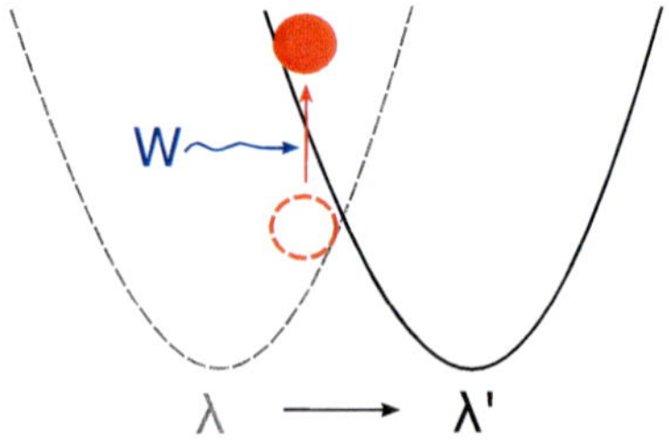

Fig. 2.5 Schematic diagram showing positive work W done by the trap potential, when its centre λ is shifted to a new position λ'

2.4.2 Discrete-Time Energy Estimate

In Sect. 2.4.1, we defined the stochastic heat and work of a fluctuating system in an external potential. We found that these energy estimates are exact and conserve the energy in the stochastic system. Although the system is fluctuating, the energy relations are exact at every instant of time. Equations (2.27) and (2.31) define the heat and work estimates in terms of the experimentally observable quantities: bead position x, control parameter λ, and external work potential $U(x, \lambda)$. These relations have led to experiments, in stochastic systems, that measure small energy flows of the order of one $k_{\mathrm{B}}T$ per relaxation time of the bead in the potential [32]. Such small energies cannot, in general, be measured directly using a physical measuring device. Since experiments measure the position of the bead at discrete time intervals, we need to formulate the discrete-time approximations of quantities that depend on continuous paths, such as work and heat.

In Eq. (2.10), we considered a bead trapped in a harmonic potential under the influence of gravity. The bead fluctuates in the harmonic potential U, exchanging energy with the environment. Following Eq. (2.31), the discrete-time work done by the trap between time steps n and $n+1$ is given by

$$\begin{aligned} W_{n+1} &= U[x_{n+1}, \lambda_{n+1}] - U[x_{n+1}, \lambda_n] \\ &= \tfrac{1}{2}\left[(x_{n+1} - \lambda_{n+1})^2 - (x_{n+1} - \lambda_n)^2\right] . \end{aligned} \tag{2.33}$$

The above equation approximates the bead position as constant while the trap position is updated. In Chap. 3, we measure the time taken to update the position of the trap and find that the trap dynamics are indeed much faster than the bead's dynamics, justifying the above approximation.

Similarly, from Eq. (2.27) the heat dissipated by the bead during one time step, using the Stratonovich convention (see[29], A. 10), is defined as

$$\Delta Q_{n+1} = \left[\frac{\left(x_{n+1} - \lambda_{n+1} - \delta_{\mathrm{g}}\right) + \left(x_n - \lambda_n - \delta_{\mathrm{g}}\right)}{2}\right](x_{n+1} - x_n) , \tag{2.34}$$

where the total external force, due to both trap and gravity, is evaluated at the centre of two neighbouring time steps.

Finally, the bead has gravitational potential energy associated with its height. The discrete-time estimate of the change in gravitational potential energy is defined as

$$F_{n+1} = \delta_{\mathrm{g}}\,(x_{n+1} - x_n) . \tag{2.35}$$

2.5 Probability Density of Bead Dynamics

In Sect. 2.3, we solved the discrete time dynamics of the bead. These equations describe the dynamics of a single trajectory of a Brownian bead. In this section, we solve the probability density function of the bead's dynamics evolving in a harmonic trap. The probability density function obeys the Fokker-Planck equation, which provides an alternate but equivalent description of the statistics of bead motion [33]. The Fokker-Planck equation is a partial differential equation that describes the time evolution of the probability density function, $P(x, t)$. Using the Fokker-Planck equation, we solve the evolution of $P(x, t)$ in a static and translating harmonic trap.

2.5.1 *Bead in a Static Harmonic Trap*

The Fokker-Planck equation of a bead in a static trap ($\lambda(t) = 0$) is given by

$$\frac{\partial P(x,t)}{\partial t} = \underbrace{\frac{\kappa}{\gamma}\frac{\partial}{\partial x} x P(x,t)}_{\text{drift}} + \underbrace{D\frac{\partial^2}{\partial x^2} P(x,t)}_{\text{diffusion}} . \tag{2.36}$$

In Eq. (2.10), the external forces acting on the bead are linear in x, and the stochastic forces are linear, additive white noise. As noted in Sect. 2.3, the probability density function of the bead's position follows a Gaussian distribution, characterized by a mean and variance [34].

To solve for the mean, we multiply both sides of Eq. 2.36 by x and integrate over all possible values of x,

$$\frac{\partial}{\partial t}\int_{-\infty}^{\infty} \mathrm{d}x\, x P(x,t) = \frac{\kappa}{\gamma}\int_{-\infty}^{\infty} \mathrm{d}x\, x \frac{\partial}{\partial x} x P(x,t) + D\int_{-\infty}^{\infty} \mathrm{d}x\, x \frac{\partial^2}{\partial x^2} P(x,t) . \tag{2.37}$$

Integrating the above equation leads to

$$\begin{aligned}\frac{\partial}{\partial t}\int_{-\infty}^{\infty} \mathrm{d}x\, x P(x,t) = \frac{\kappa}{\gamma}\Bigg[&\int_{-\infty}^{\infty} \mathrm{d}x\, x P(x,t) + x^2 P(x,t)\Big|_{-\infty}^{+\infty} \\ &-2\int_{-\infty}^{\infty} \mathrm{d}x\, x P(x,t)\Bigg] \\ &+ D\left[x\frac{\partial}{\partial x} P(x,t)\Big|_{-\infty}^{+\infty} - P(x,t)\Big|_{-\infty}^{+\infty}\right] .\end{aligned} \tag{2.38}$$

Using the fact that the probability of finding the bead at infinity vanishes, we find the conditions $P(x,t)|_{\pm\infty} = 0$ and $\left.\frac{\partial P(x,t)}{\partial x}\right|_{\pm\infty} = 0$. Substituting the boundary terms and using the definition $\langle x(t)\rangle \equiv \int_{-\infty}^{\infty} \mathrm{d}x\, x P(x,t)$, we find that the mean of the probability density function satisfies

$$\frac{\mathrm{d}}{\mathrm{d}t}\langle x(t)\rangle = -\frac{\kappa}{\gamma}\langle x(t)\rangle \,. \tag{2.39}$$

Using the initial condition, $x(t=0) = x_0$, we find a solution

$$\langle x(t)\rangle = x_0 \mathrm{e}^{-\kappa t/\gamma} \,. \tag{2.40}$$

Next, we solve for the second moment of the distribution. We multiply Eq. (2.36) by x^2 and again integrate x in the range $(-\infty, \infty)$,

$$\frac{\partial}{\partial t}\int_{-\infty}^{\infty} \mathrm{d}x\, x^2 P(x,t) = \frac{\kappa}{\gamma}\int_{-\infty}^{\infty} \mathrm{d}x\, x^2 \frac{\partial}{\partial x} x P(x,t) + D\int_{-\infty}^{\infty} \mathrm{d}x\, x^2 \frac{\partial^2}{\partial x^2} P(x,t) \,. \tag{2.41}$$

Following a procedure similar to that used to calculate the equation of motion of the mean, and using the condition $\int_{-\infty}^{\infty} \mathrm{d}x\, P(x,t) = 1$, we find

$$\frac{\mathrm{d}}{\mathrm{d}t}\left\langle x^2(t)\right\rangle = -2\frac{\kappa}{\gamma}\left\langle x^2(t)\right\rangle + 2D \,. \tag{2.42}$$

If the particle position is known exactly at time $t = 0$, then $\left\langle x^2(t=0)\right\rangle = 0$, and

$$\left\langle x^2(t)\right\rangle = \frac{D\gamma}{\kappa}\left[1 - \exp(-\frac{2\kappa t}{\gamma})\right] \,. \tag{2.43}$$

In the long-time limit, the mean μ $(= \langle x(t \to \infty)\rangle)$ and variance σ^2 $\left(=\left\langle x^2\right\rangle - \langle x\rangle^2\right)$ of the probability density function are

$$\mu = 0 \tag{2.44a}$$

$$\sigma^2 = \frac{D\gamma}{\kappa} \,. \tag{2.44b}$$

The probability density function thus evolves from an off-centre delta function to a centred Gaussian distribution with variance given by Eq. (2.44b), as shown in Fig. 2.6. The long-time stationary distribution corresponds to thermal equilibrium, and the distribution is known as a Boltzmann distribution. Thus, Fig. 2.6 also represents the dynamics of thermal relaxation.

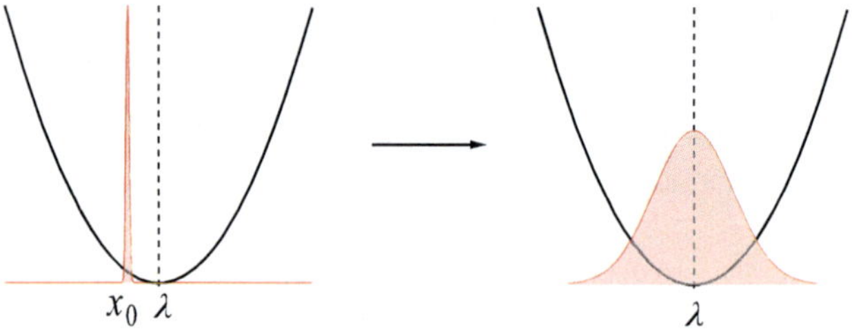

Fig. 2.6 Probability density function of a bead whose position is known exactly at $t = 0$ evolves in a static trap from an off-centre delta function to a centred Gaussian distribution

2.5.2 *Bead in a Translating Trap*

Here, we solve for the distribution of bead positions, when the trap is shifted at a constant velocity $\lambda(t) = vt$ [34]. The Langevin equation is given by

$$\dot{x} = -\frac{\kappa}{\gamma}(x - vt) + \sqrt{2D}\,\xi(t)\,. \tag{2.45}$$

By writing the equation in the trap's frame of reference using $y = x - vt$, we find

$$\dot{y} = -\frac{\kappa}{\gamma}y - v + \sqrt{2D}\,\xi(t)\,. \tag{2.46}$$

The corresponding Fokker-Planck equation is given by

$$\frac{\partial P(y,t)}{\partial t} = \frac{\kappa}{\gamma}\frac{\partial}{\partial y}yP(y,t) + v\frac{\partial}{\partial y}P(y,t) + D\frac{\partial^2}{\partial y^2}P(y,t)\,. \tag{2.47}$$

Multiplying the Eq. (2.47) by y and integrating y over the range $(-\infty, \infty)$, we find

$$\frac{\mathrm{d}}{\mathrm{d}t}\langle y(t)\rangle = -\frac{\kappa}{\gamma}\langle y(t)\rangle - v\,. \tag{2.48}$$

The solution to Eq. (2.48) is given by

$$\langle y(t)\rangle = -\frac{\gamma v}{\kappa} + \left(y_0 + \frac{\gamma u}{\kappa}\right)\mathrm{e}^{-(\kappa/\gamma)t}\,. \tag{2.49}$$

Similarly, the second moment is obtained by multiplying the equation by y^2 and integrating the equation over all possible values of y, giving

$$\frac{\mathrm{d}}{\mathrm{d}t}\left\langle y^2(t)\right\rangle = -2\frac{\kappa}{\gamma}\left\langle y^2(t)\right\rangle + 2D\,. \tag{2.50}$$

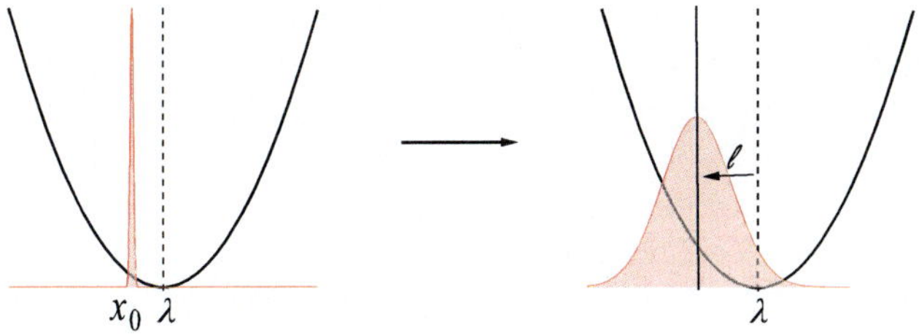

Fig. 2.7 Schematic diagram of probability density function of a bead, initialized to x_0, in a harmonic trap moving at a velocity v. The shift in the mean of the distribution ℓ from the centre of the trap is given by Eq. (2.52a)

The solution to the above equation is

$$\left\langle y^2(t) \right\rangle = \frac{D\gamma}{\kappa}\left[1 - \exp(-\frac{2\kappa t}{\gamma})\right], \tag{2.51}$$

where the initial condition $\left\langle y^2(t=0) \right\rangle = 0$. In the long-time limit, the mean ℓ and variance σ^2 of the distribution are

$$\ell = -\frac{\gamma v}{\kappa} \tag{2.52a}$$

$$\sigma^2 = \frac{D\gamma}{\kappa}. \tag{2.52b}$$

Comparing the above steady-state solution to Eq. (2.44b), we note that the variance of the distribution remains the same, while the mean of the distribution is shifted from the centre of the trap by an amount that depends on the velocity, as shown in Fig. 2.7. Compare to Fig. 2.6, where there is no long-time shift.

2.6 Hydrodynamics of a Moving Bead

In the previous section, we looked at the probability density function for the bead position in static and translating traps. In this section, we study the hydrodynamic flows generated by a translating bead in the surrounding fluid medium. Consider a microscopic bead moving through a fluid at a constant velocity v. The velocity field of the fluid surrounding the bead is described by the Navier-Stokes equation [9],

$$\rho\left(\frac{\partial u}{\partial t} + u \cdot \nabla u\right) = \eta \nabla^2 u - \nabla p, \tag{2.53}$$

where u is the three-dimensional vector describing the velocity of the fluid at position x and time t, ρ the fluid density, η the fluid viscosity, and p the pressure.

The first term on the left is the acceleration of the fluid at position x, and the second term is the change in the fluid velocity due to transport of fluid to a different position by convection. For an incompressible Newtonian fluid, the velocity field also follows the continuity equation $\nabla \cdot u = 0$.

To solve the above equation for the velocity of the fluid, we need to specify the boundary conditions. For a bead moving in a fluid, we assume that the flow field is undisturbed at infinity. At the surface of the bead, we apply no-slip, or "stick", boundary conditions; i.e., the velocity of the fluid equals the bead's velocity at the bead surface. At steady state, the acceleration term $\partial u/\partial t$ can be dropped. When scaled by a characteristic length ℓ_0 and velocity u_0, the non-dimensional Navier-Stokes equation is given by

$$\mathrm{Re}\, u' \cdot \nabla u' = -\nabla' p' + \nabla'^2 u', \tag{2.54}$$

where Re is the *Reynolds number*, given by $\mathrm{Re} = \frac{\ell_0 u_0 \rho}{\eta}$. The fluid shows different flow characteristics at different Reynolds numbers. For $\mathrm{Re} \gg 1$ the fluid flow is turbulent, whereas for $\mathrm{Re} \ll 1$, the flow is laminar [9]. In the low-Reynolds-number limit, the Navier-Stokes equation simplifies to the Stokes equation

$$\eta \nabla^2 u = \nabla p\,. \tag{2.55}$$

Solving the Stokes equation with the no-slip boundary condition at the surface of the bead, one can show that the velocity field of the fluid around a bead translating at velocity v is [35]

$$u_R = \frac{1}{2} v \cos(\phi) \left(\frac{r}{R}\right)^2 \left(\frac{3R}{r} - \frac{r}{R}\right) \tag{2.56a}$$

$$u_\phi = -\frac{1}{4} v \sin(\phi) \left(\frac{r}{R}\right) \left(\left(\frac{r}{R}\right)^2 + 3\right) \tag{2.56b}$$

where r is the radius of the bead, ϕ is the angle with respect to the direction of the bead's motion, and R is the radial distance from the centre of the bead. The flow fields are shown in Fig. 2.8.

Consider another bead in the medium at a distance R from the translating bead. Because of the fluid flow produced by the first bead, the second bead experiences a hydrodynamic force F_{hydro}. We use the second bead as a probe to measure the hydrodynamic flow generated by the first (driving) bead. The equation of motion of the second bead follows

$$-\gamma \dot{x}_p = F_{\mathrm{hydro}} + \xi_p(t)\,, \tag{2.57}$$

where ξ_p is the thermal force and x_p is the position of the probe bead.

The hydrodynamic coupling of the beads is experimentally measured using a dual-trap apparatus. Let us consider two beads of equal radii r in separate

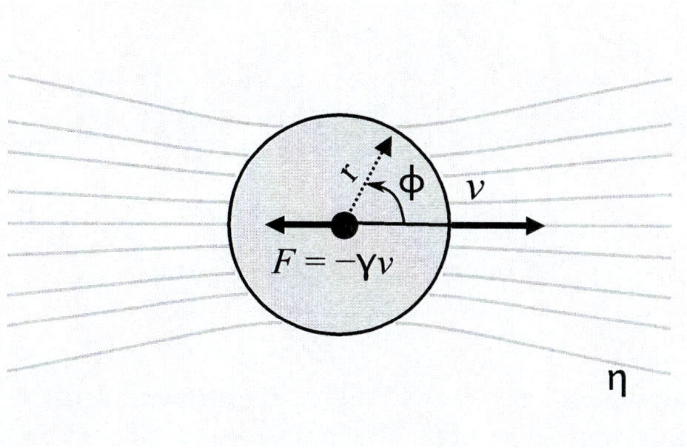

Fig. 2.8 The hydrodynamic flow field generated by a bead of radius r translating at a constant velocity v in a fluid of viscosity η, at low Re

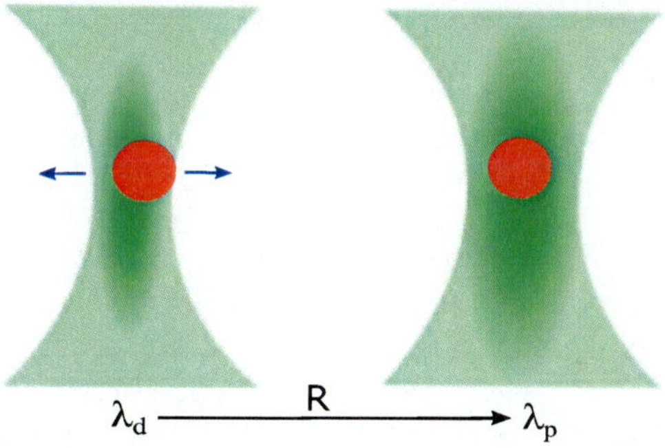

Fig. 2.9 Schematic diagram of dual tweezers whose centres at λ_d and λ_p are separated by a distance R. The blue arrows indicate the direction of motion of the trap

harmonic traps, with the trap centres separated by a distance R, as shown in Fig. 2.9. The position of the centre of the first trap is modulated sinusoidally. For a slow modulation frequency $\omega_d \ll 1/\tau_r$, where τ_r is the relaxation time of the bead in the left trap, the driving bead follows the trap centre and creates quasistatic flow fields. We can measure the response of the probe bead due to the hydrodynamic coupling by using the cross-correlation spectrum. Neglecting the stochastic forcing terms, we can write the equations of motion as

$$\dot{x}_d = -\kappa_d\, \mu(x_d - \lambda_d) - \kappa_p\, \mu'(x_p - \lambda_p) \tag{2.58a}$$

$$\dot{x}_p = -\kappa_p\, \mu(x_p - \lambda_p) - \kappa_d\, \mu'(x_d - \lambda_d) \tag{2.58b}$$

where x_d is the position of the driving bead, x_p the position of the probe bead, $\mu = 1/6\pi\eta a$ the self-mobility and $\mu' = \frac{1}{8\pi\eta R}(2 - \frac{4a^2}{3R^2})$ the cross-mobility due to the hydrodynamic coupling [36]. The cross-correlation spectrum is calculated by taking the Fourier transform of the equation

$$\frac{\mathrm{d}X}{\mathrm{d}t} = -MX + M\Lambda \tag{2.59}$$

where $M = \begin{bmatrix} \mu\,\kappa_d & \mu'\kappa_p \\ \mu'\kappa_d & \mu\,\kappa_p \end{bmatrix}$, $X = \begin{bmatrix} x_d \\ x_p \end{bmatrix}$, and $\Lambda = \begin{bmatrix} \lambda_d \\ \lambda_p \end{bmatrix}$. The Fourier transform of Eq. (2.59) is

$$X(\omega) = [M - i\omega I]^{-1}\ M\ \Lambda(\omega) \tag{2.60a}$$

$$= \chi(\omega)\ \Lambda(\omega), \tag{2.60b}$$

where χ is the response matrix and I the 2×2 identity matrix. For a sinusoidal modulation of the trap centre of the driving bead $\lambda_d(t) = A\sin(\omega_0 t)$, the Fourier transform is $\lambda_d(\omega) = \frac{A}{2}\,(\delta(\omega - \omega_0) + \delta(\omega + \omega_0))$. The response of the probe bead because of the driving bead modulation is given by $x_p(\omega) = \chi_{21}\lambda_d(\omega)$, where χ_{12} denotes the cross coupling term in the response matrix χ. The amplitude of the cross-correlation response, $C = |\chi_{21}|A$, of the probe bead, at the driving frequency ω_0, is [36]

$$C = \frac{\omega_0\mu'\,k_d\,A}{\sqrt{(\mathrm{Det}\,M - \omega_0^2)^2 + \omega_0^2(\mathrm{Tr}\,M)^2}}, \tag{2.61}$$

where Det M and Tr M are the determinant and trace of the matrix M. In this thesis, we measure only the amplitude response C of the probe bead at low sampling rate ($t_s \gg \tau_r$). However, the phase and amplitude response of the probe bead for static [37] and oscillating [38] driving beads have been previously measured at fast sampling rates ($t_s \ll \tau_r$).

In Sect. 3.7, we discuss a dual-trap apparatus that we built to measure the cross-correlation between two trapped beads, in close vicinity, arising from hydrodynamic coupling. In Chap. 8 and Appendix B, we use this technique to study hydrodynamic coupling in the presence of feedback.

2.7 Summary

In this Chapter, we discussed the dynamics of a bead in a fluid medium described by the BBO equation. We could simplify the BBO equation to an overdamped Langevin equation since measurement time scales are slower than those of the hydrodynamic memory and inertial effects. As the positions are measured at discrete time intervals, we calculated the discrete-time equation of motion of the bead. Then we discussed stochastic thermodynamics, an approach to calculating work and heat for the bead as it moves in a potential and undergoes various dynamical protocols. Finally, we discussed the distribution of a bead translating under the influence of a moving trap at constant velocity and calculated the corresponding hydrodynamic flow generated by the bead. We use these theoretical models to describe the experimental data presented in this thesis.

References

1. R. Brown, A brief account of microscopical observations made in the months of June, July and August 1827, on the particles contained in the pollen of plants; and on the general existence of active molecules in organic and inorganic bodies. Philos. Mag. **4**, 161–173 (1828)
2. T. Li, M.G. Raizen, Brownian motion at short time scales. Ann. Phys. **525**, 281–295 (2013)
3. K. Jacobs, *Stochastic Processes for Physicists: Understanding Noisy Systems* (Cambridge University Press, Cambridge, 2010)
4. A.B. Basset, III, On the motion of a sphere in a viscous liquid. Philos. Trans. R. Soc. A **179**, 43–63 (1888)
5. J. Boussinesq, Sur la résistance qu'oppose un fluide indefini en repos, sans pesanteur, au mouvement varié d'une sphère solide qu'il mouille sur toute sa surface, quand les vitesses restent bien continues et assez faibles pour que leurs carrés et produits soient négligeables. C R Acad. Sc. Paris **100**, 935–937 (1885)
6. C.W. Oseen, *Neuere methoden und ergebnisse in der hydrodynamik* (Akademische Verlagsgesellschaft mb H., Leipzig, 1927)
7. S.L. Seyler, S. Pressé, Surmounting potential barriers: Hydrodynamic memory hedges against thermal fluctuations in particle transport. Chem. Phys. **153**, 041102 (2020).
8. T.S. Chow, J.J. Hermans, Effect of inertia on the Brownian motion of rigid particles in a viscous fluid. J. Chem. Phys. **56**, 3150–3154 (1972)
9. D.J. Tritton, *Physical Fluid Dynamics* (Springer Science & Business Media, New York, 2012)
10. M. Parmar, A. Haselbacher, S. Balachandar, Generalized Basset-Boussinesq-Oseen equation for unsteady forces on a sphere in a compressible flow. Phys. Rev. Lett. **106**, 084501 (2011)
11. J. Tóthová, V. Lisý, A note on the fluctuation-dissipation relation for the generalized Langevin equation with hydrodynamic backflow. Phys. Lett. A **380**, 2561–2564 (2016)
12. K. Sekimoto, *Stochastic Energetics* (Springer, Berlin, 2010)
13. G.E. Uhlenbeck, L.S. Ornstein, On the theory of the Brownian motion. Phys. Rev. **36**, 823 (1930)
14. J.S. Turner, *Buoyancy Effects in Fluids* (Cambridge University Press, Cambridge, 1979)
15. A. Ashkin, Acceleration and trapping of particles by radiation pressure. Phys. Rev. Lett. **24**, 156 (1970)
16. A. Ashkin, Atomic-beam deflection by resonance-radiation pressure. Phys. Rev. Lett. **25**, 1321 (1970)
17. A. Ashkin, J.M. Dziedzic, J.E. Bjorkholm, S. Chu, Observation of a single-beam gradient force optical trap for dielectric particles. Opt. Lett. **11**, 288–290 (1986)
18. A. Ashkin, Optical trapping and manipulation of neutral particles using lasers. Proc. Natl. Acad. Sci. U.S.A **94**, 4853–4860 (1997)
19. A. Ashkin, Trapping of atoms by resonance radiation pressure. Phys. Rev. Lett. **40**, 729 (1978)
20. A. Ashkin, J.M. Dziedzic, T. Yamane, Optical trapping and manipulation of single cells using infrared laser beams. Nature **330**, 769–771 (1987)
21. A. Kumar, J. Bechhoefer, Nanoscale virtual potentials using optical tweezers. Appl. Phys. Lett. **113**, 183702 (2018)
22. A. Ashkin, J.M. Dziedzic, J.E. Bjorkholm, S. Chu, Observation of a single-beam gradient force optical trap for dielectric particles. Opt. Lett. **11**, 288–290 (1986)
23. A. Ashkin, Forces of a single-beam gradient laser trap on a dielectric sphere in the ray optics regime. Biophys. J. **61**, 569–582 (1992)

24. G. Pesce, P.H. Jones, O.M. Maragò, G. Volpe, Optical tweezers: theory and practice. Eur. Phys. J. Plus **135**, 1–38 (2020)
25. H.C. Hulst, H.C. van de Hulst, *Light Scattering by Small Particles* (Courier Corporation, Chelmsford, 1981)
26. G. Gouesbet, G. Gréhan, *Generalized Lorenz-Mie Theories* (Springer, Berlin, 2011)
27. B.T. Draine, J. Goodman, Beyond Clausius-Mossotti-Wave propagation on a polarizable point lattice and the discrete dipole approximation. Astrophys. J. **405**, 685–697 (1993)
28. Y. Roichman, B. Sun, A. Stolarski, D.G. Grier, Influence of nonconservative optical forces on the dynamics of optically trapped colloidal spheres: the fountain of probability. Phys. Rev. Lett. **101**, 128301 (2008)
29. L. Peliti, S. Pigolotti, *Stochastic Thermodynamics: An Introduction* (Princeton University Press, Princeton, 2021)
30. B. Lau, O. Kedem, J. Schwabacher, D. Kwasnieski, E.A. Weiss, An introduction to ratchets in chemistry and biology. Mater. Horiz. **4**, 310–318 (2017)
31. K. Sekimoto, Langevin equation and thermodynamics. Prog. Theor. Phys. **130**, 17–27 (1998)
32. S. Ciliberto, Experiments in stochastic thermodynamics: short history and perspectives. Phys. Rev. X **7**, 021051 (2017)
33. H. Risken, *The Fokker-Planck Equation*, 2nd. edn., Chap. 8.1 (Springer, Berlin/Heidelberg/New York, 1996)
34. O. Mazonka, C. Jarzynski, Exactly solvable model illustrating far-from-equilibrium predictions. Preprint (1999). arXiv:cond-mat/9912121
35. T.G.M. Van de Ven, *Colloidal Hydrodynamics* (Academic Press, Cambridge, 1989)
36. S. Paul, R. Kumar, A. Banerjee, Two-point active microrheology in a viscous medium exploiting a motional resonance excited in dual-trap optical tweezers. Phys. Rev. E **97**, 042606 (2018)
37. J.-C. Meiners, S.R. Quake, Direct measurement of hydrodynamic cross correlations between two particles in an external potential. Phys. Rev. Lett. **82**, 2211 (1999)
38. S. Paul, et al., Direct verification of the fluctuation-dissipation relation in viscously coupled oscillators. Phys. Rev. E **96**, 050102 (2017)

Chapter 3
Experimental Apparatus

In this chapter, we discuss the experimental apparatus that is used to build, control, and characterize the information engine. We discuss the calibration routines that convert the (nonlinear) voltage signal from the detector into a position of the bead in physical units. Then we discuss a method to measure the optical trap stiffness, bead's diffusion constant, and the measurement noise of the position detector. We also present a method we developed to control the strength of the measurement noise. Then we discuss an optimal-filtering technique that improves the estimate of the bead's position in the presence of measurement noise. We implement the real-time position estimate and benchmark its performance. Next, we modify the experimental apparatus to integrate a dual optical trap system and trap one bead in each trap. Finally, we modify the sample chamber design to introduce electrodes that can apply external forces on the bead.

Contributions to the chapter: Avinash Kumar and John Bechhoefer developed the first version of the optical feedback tweezers. Luis Reinalter, a summer student in the lab, assembled a new optical feedback tweezers setup based on that first version. To make the new setup useful for creating and studying information engines, I reduced the electrical noise, identified and replaced faulty equipment, improved the apparatus design, and benchmarked the performance of the new apparatus. Next, I automated the nonlinear calibration routine of the position detector using LabVIEW. I modified the apparatus to add the measurement noise control and dual-trap system. I collaborated with Momčilo Gavrilov to build a modified sample chamber to add external electric forces. I performed the experiments and analyzed the data presented in this chapter.

T. K. Saha, *Information-Powered Engines*, Springer Theses,
https://doi.org/10.1007/978-3-031-49121-4_3

3.1 Optical Feedback Tweezers

The experimental apparatus consists of a bead in an optical trap whose position is controlled by a feedback loop. The optical trap is created by tightly focusing a 532-nm green laser beam (Cobolt, Samba 1.5 W) using a microscope objective, as shown in Fig. 3.1. The trap stiffness depends on the intensity gradient of the focused beam. A high-intensity gradient is achieved by using a high-numerical-aperture microscope objective. We chose a water-immersion microscope objective (MO1, Olympus NA = 1.2) as the trapping objective. Oil-immersion objectives have a higher numerical aperture and are designed to minimize aberrations at the glass-water interface created by the coverslip. By contrast, water-immersion objectives are designed to produce only a small amount of spherical aberration when focused (relatively) deep in a water sample. They thus maintain a small spot size (strong trap) when trapping particles tens of microns deep into the water medium. In our experiments, we want to trap beads far from any surface so that properties such as the diffusion constant are, to a good approximation, independent of the distance to the nearest surface.

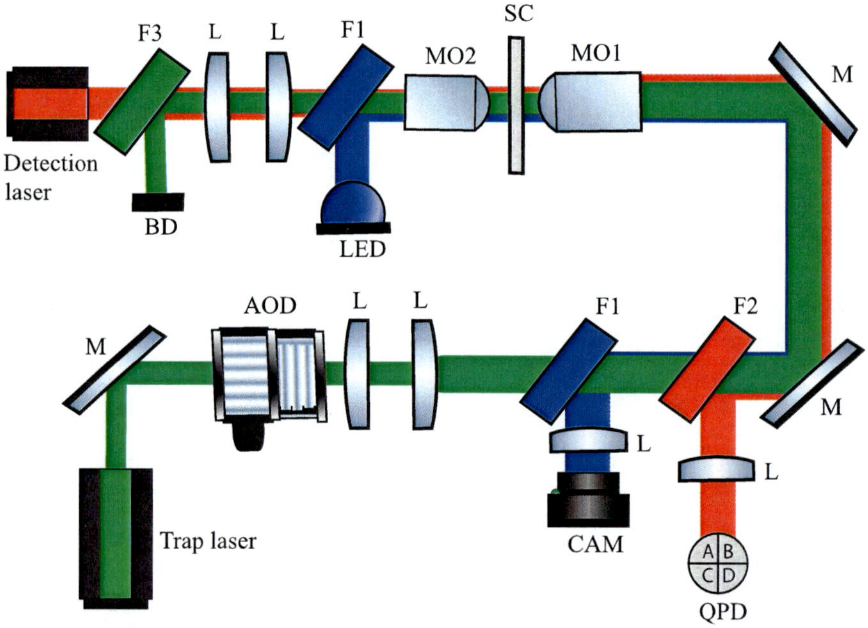

Fig. 3.1 Schematic diagram of the experimental apparatus. *M* mirror, *AOD* acousto-optic deflector, *L* lens, *F1* blue dichroic filter, *CAM* camera, *F2* red dichroic filter, *MO1* trapping microscope objective, *MO2* detection microscope objective, *SC* sample chamber, *F3* green dichroic filter, *BD* beam dump, *QPD* quadrant photo-diode

The position of the trapped bead is measured using the forward-scattered light from a counter-propagating 660-nm red (Cobolt, Flamenco 100 mW) detection laser. The detection laser is loosely focused on the bead using another microscope objective (MO2, Nikon, 0.5 NA). The detection plane is aligned with the trapping plane using a two-lens (focal length = 30 mm) telescope system. The forward-scattered detection light is collected by the trapping microscope objective and isolated from the trapping beam's path using a dichroic mirror (F2). It is then imaged onto a quadrant photodiode (QPD, First Sensor, QP50-6-18u-SD2), using a convex lens (focal length = 175 mm). The forward-scattered light from the bead creates an interference pattern that shifts with the bead's position. The changes in the detection-laser intensity pattern contain the information of the bead's position in the detection plane. The QPD measures the laser intensity falling on the different quadrants and outputs a voltage signal. This voltage signal is then converted to a position signal for the bead using a calibration routine discussed in Sect. 3.2.

The information engine operates by applying feedback forces based on the bead's position. The feedback forces are generated by changing the position of the trap. The trap's position is shifted using two acousto-optic deflectors (AODs, DTSXY-250-532, AA Opto-Electronic), one each for the x and y axes, as shown in Fig. 3.1. The AOD changes the position of the trap (with a rise time $\approx 1\,\mu$s) quickly compared to the relaxation time of the bead in the trap τ_r ($\approx$ 1 ms), which makes it a good candidate to build optical feedback traps [1]. In Sect. 3.3, we discuss the working principle of the AOD and characterize its performance more precisely.

The feedback forces applied on the bead are calculated using a field programmable gate array (FPGA, National Instruments PCIe 7857). It can implement a sequence of computations at well-defined intervals of time, with a clock rate of 80 MHz and a timing jitter 250 ps. The FPGA interacts with the sensors and actuators to implement the feedback control. The FPGA converts the QPD voltage into position using the pre-recorded nonlinear calibration fit parameters, found using the routine discussed in Sect. 3.2. Based on the feedback rule, the FPGA calculates the updated trap position. The calculated trap position is converted into an AOD voltage by the FPGA (using the calibration method discussed in Sect. 3.2), and is sent to an analog controller (a voltage-controlled oscillator (VCO) circuit, AA Opto-Electronic DRFA10Y-B-0-50.110). The analog controller generates the corresponding control signal for the AOD. The AOD then updates the position of the trap. The position of the trap is updated 10 μs after the output voltage is sent by the FPGA (see Sect. 3.3). After sending the output AOD voltage, the FPGA stores the data to the on-board memory during the next 10 μs while the AOD updates the trap position, as shown in Fig. 3.2. The feedback loop runs at a rate of 50 kHz.

The final section of the experimental apparatus is the camera imaging system. We introduced a blue LED using a high-pass dichroic filter (F1) into the laser path, to illuminate the bead. The illumination light collected by the objective MO1 is imaged on a camera (BASLER, Ace acA800-510 μm), using a lens of focal length 175 mm.

The sample chamber is made using a glass slide and a coverslip, separated using 100-μm spacer wires on two sides. The sides of the sample chamber are sealed using

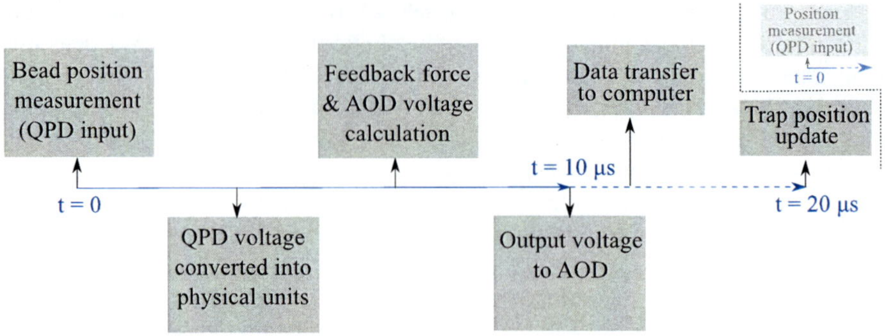

Fig. 3.2 Structure of the feedback loop diagram implemented on the FPGA (field programmable gate array). *QPD* quadrant photodiode, *AOD* acousto-optic deflector

nail polish, leaving small gaps on opposite ends of the chamber without the wires. The sample chamber is filled with $\approx 45\ \mu$L of the bead solution. The bead solution is made using 0.05% NaN_3 w/v solution in 50 mL deionized water, to inhibit bacterial growth. We then add 2 μL of 3-μm beads (Sigma-Aldrich, 5% solid concentration). For different bead diameters, the concentration was adjusted to obtain a number density of about one bead per camera field of view ($64 \times 48\ \mu$m). The depth of field of the camera image is approximately 1 μm.

The sample chamber is mounted to a home-built sample-chamber holder using magnets. The sample-chamber holder is attached to a linear translation stage (Newport, DS25–XYZ). A bead is trapped by bringing it close to the trap using the translation stage. The camera is used to visualize the bead in the trapping plane and to approximately align the laser beams. The imaging system is also used to calibrate the AOD and QPD voltages (Sect. 3.2).

3.2 Bead and Trap Position Calibration

The bead and trap positions are calibrated using the pixels of the camera, which have a known size. To calibrate the camera, we take an image of the calibration scale, as shown in Fig. 3.3. The image-processing software ImageJ (https://imagej.nih.gov/ij/download.html) is used to convert the pixels into real units. The calibration factor was measured to be 82.5 ± 0.1 nm/pixel.

The trap position is calibrated by imaging a trapped bead. The position of the trap is shifted using the AOD. The input voltage range of the AOD is 0–10 V. The control voltage of the AOD is changed in the range 4.8–5.2 V, in steps of 0.05 V. The corresponding position of the bead is recorded using the camera, using a single image for each AOD voltage. Since the relaxation time of the bead is usually 1 ms, we wait for $\approx$ 1 s (1000 relaxation times) before recording the bead position after

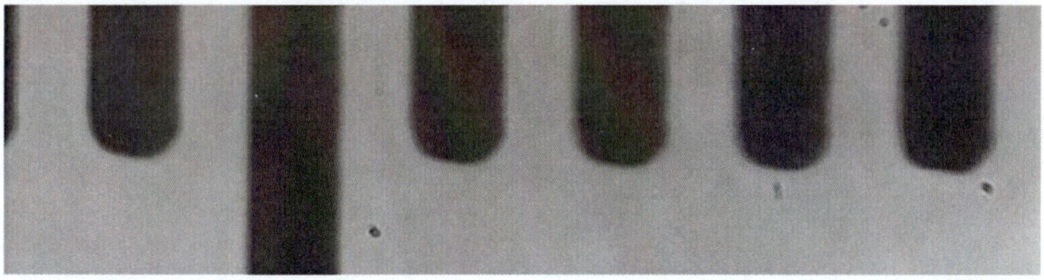

Fig. 3.3 Calibration scale used to convert pixels to real units. The minor ticks are 10 μm apart

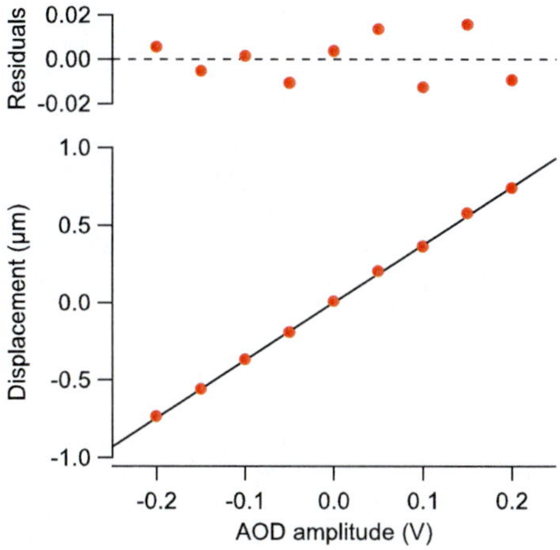

Fig. 3.4 Calibration of the AOD by measuring the displacement of the bead for different AOD input voltages. The input range is ± 0.2 V, centred on a DC offset of 5.0 V

changing the AOD voltage. The protocol leads to a good approximation of the trap centre, as indicated by the residuals ($\approx$ 10 nm), which are small compared to the total detection region ($\approx$ 1400 nm). The residuals are comparable to the standard deviation of the bead in the trap ($\sigma \approx 20$ nm), and can be reduced by the making multiple measurements. Finally, the bead displacement vs. AOD voltage change is fit to a line to measure the AOD calibration factor, as shown in Fig. 3.4. The calibration factor of the AOD was measured to be 3.73 ± 0.02 μm/V.

The QPD is calibrated using an automated least-squares nonlinear fitting routine available in LabVIEW. The position of the bead is quasistatically modulated, as a triangle wave of amplitude 0.2 V centred at 5 V, using the AOD. The corresponding signal of the bead position is recorded, as shown in Fig. 3.5a. The QPD response is fit to a fifth-order polynomial. Using the polynomial coefficients, we plot the linearized response of the QPD in Fig. 3.5b, shown as the red curve. The displacements of the bead as a function of the AOD voltage are fit to a line (black curve), as shown in Fig. 3.5b with the corresponding residuals of the linear fit. From the smoothed curve of the residuals, we estimate the systematic errors in the calibration to be $\lesssim$ 10 nm over the range of 1420 nm (i.e., < 1% of the full-range scale).

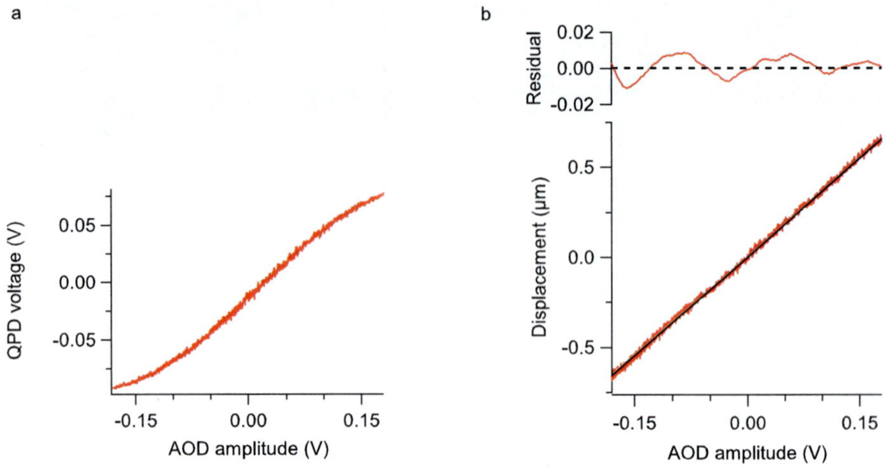

Fig. 3.5 Calibration of the QPD. (**a**) QPD signal of a bead modulated as a triangle wave by the AOD of amplitude 0.2 V centred at 5 V. (**b**) Inferred bead position after nonlinear calibration of the QPD signal. The black solid line denotes the linear fit of the bead displacement vs AOD voltage. The corresponding residuals of the fit are presented as a smoothed curve (using a Savitzky–Golay filter, 2nd order averaged over 5001 points)

Fig. 3.6 Schematic diagram showing the working of the AOD. The red arrow represents the direction of the travelling wave

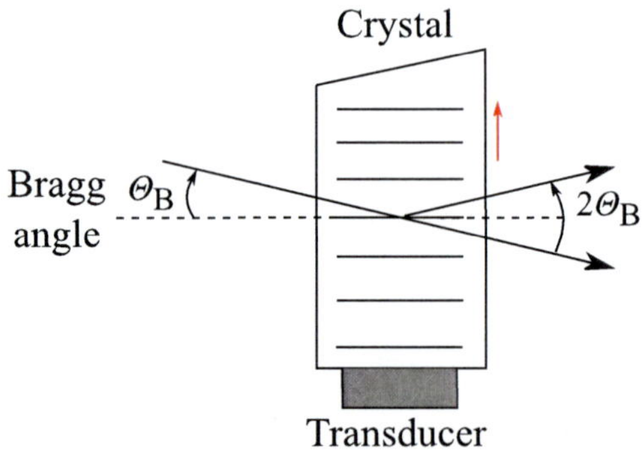

3.3 Acousto-Optic Deflector

The AOD consists of a crystal through which a propagating sound wave periodically modulates the index of refraction, creating a refractive-index grating. The frequency and amplitude of the sound wave is controlled using a voltage-controlled oscillator (VCO) circuit, and the signal is amplified using an RF amplifier (AA Opto-Electronic, AMPA-B-30). The laser beam, when incident at the Bragg angle, generates a first-order diffraction pattern without any higher-order diffraction patterns. The angle that the first order makes with respect to the unscattered beam, $2\theta_B$, depends on the frequency of the sound wave (Fig. 3.6).

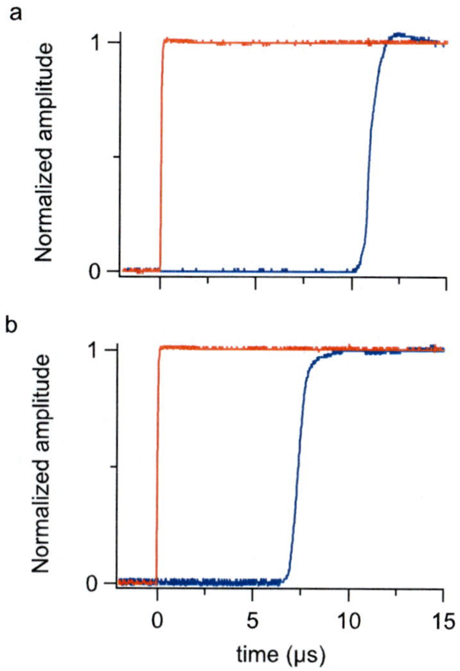

Fig. 3.7 AOD response delay when the frequency modulation is controlled by (**a**) an analog controller, and (**b**) a digital controller

The sound wave propagates in the TeO_2 crystal at a speed ≈ 660 m/s. The wave is generated by a transducer attached to one end of the crystal and absorbed by the other end, which is also angled to further reduce the amplitude of reflected waves. The time for the sound wave to cross a laser beam of diameter 1 mm is ≈ 1 μs. For a 1-cm-long crystal, the sound wave takes ≈ 6.5 μs to propagate from the transducer to the edge of the beam, closer to the transducer. The physical distance between the transducer and the laser beam leads to a delay in the trap update. Figure 3.7a shows the AOD response (red) measured using a photodetector to a step-function generated by the FPGA ($t = 0$ denotes the time at which the FPGA updates the control signal). The total delay of the AOD is ≈ 10 μs, which is a sum of the delay from the sound wave propagation in the crystal and the delay from the analog-to-digital converter in the analog VCO controller.

To measure the delay arising from the analog VCO controller, we replace the analog controller with a direct-digital synthesizer (DDS, AA opto-electronics, DDSPA-B431b-0). Replacing the analog with the digital controller and keeping the geometry of beam and transducer the same reduced the total delay to $\approx$ 7.5 μs, as shown in Fig. 3.7b. The analog controller thus accounts for ≈ 2.5 μs of the total AOD delay. The remaining delay arises from the physical system: the time it takes for the sound wave to propagate in the crystal. The minimum delay that could be achieved by positioning the beam near the edge of the AOD crystal was ≈ 5 μs, which can ideally double the feedback loop rate. Because the configuration for minimum delay requires more alignment and is more sensitive to small equipment displacements, we did not use the minimum-delay configuration. Note that the experiments presented in this thesis were performed using the analog VCO controller.

3.3.1 Feedback Delay

Accurate energy measurements of stochastic systems require precise feedback-loop timing. After the bead's position is measured by the QPD, the FPGA takes $< 10\,\mu s$ to compute and output the feedback AOD voltage, and the AOD takes up to $10\,\mu s$ more to update the trap position. The total feedback latency (informally called feedback delay, t_d) is thus $< 20\,\mu s$ from the position measurement to the trap update. To simplify the feedback loop structure, we set the sampling time $t_s = 20\,\mu s$. The sampling time is set to be greater than the total feedback latency so that any small fluctuation in the loop time arising from changes in the LabVIEW code does not affect the feedback-loop time. Setting $t_s = t_d$ simplifies both the feedback algorithm and the energy calculations used later in this thesis [2]. For example, the position measurement occurs at the same time the trap is moved (in response to the previous measurement).

3.4 Experimental Parameter Estimates

The trap stiffness κ and the diffusion constant D of the bead are calculated by fitting the power spectrum of the bead position. The discrete-time dynamics of the bead is given by Eq. (2.20). Converting the position and time back into real units, in order to measure the parameters, the equation of motion is given by

$$x_{n+1} = e^{-t_s/\tau_r} x_n + \sigma\sqrt{1 - e^{-2t_s/\tau_r}}\, \xi_n\,, \tag{3.1}$$

where $\delta_g = 0$, and the positions are measured in the trap-centre's frame of reference. The sampling frequency is defined as $f_s = 1/t_s = 2 f_N$, where f_N is the Nyquist frequency. It is the highest frequency, for a given sampling frequency, for which the true system dynamics can be reconstructed without aliasing [3]. Frequencies above the Nyquist frequency have contributions from lower harmonics, a phenomenon known as aliasing. Using the relations $\tau_r = 1/2\pi f_c$ and $\sigma^2 = k_B T/\kappa = D/2\pi f_c$, we find

$$\begin{aligned} x_{n+1} &= e^{-\pi f_c/f_N}\, x_n + \sqrt{\frac{D}{2\pi f_c}}\sqrt{1 - e^{-2\pi f_c/f_N}}\, \xi_n \\ &= c\, x_n + \left(\frac{(1-c^2)D}{2\pi f_c}\right)^{1/2} \xi_n \\ &= c\, x_n + \Delta x\, \xi_n\,, \end{aligned} \tag{3.2}$$

where D $(= k_BT/\gamma)$ is the diffusion constant of the bead, and $c = \mathrm{e}^{-\pi f_c/f_N}$. The discrete-time Fourier transform of the position x is given by

$$\tilde{x}_k = t_s \sum_{n=1}^{N} \mathrm{e}^{i2\pi f_k t_n} x_n = t_s \sum_{n=1}^{N} \mathrm{e}^{i2\pi kn/N} x_j \,, \tag{3.3}$$

where $t_n = nt_s$, N is the total number of measurements, $T = Nt_s$ is the total time, and $k = -\frac{N}{2}+1, \ldots, \frac{N}{2}$. The Fourier transform of the noise ξ has respective mean and variance

$$\left\langle \tilde{\xi}_k \right\rangle = 0, \quad \text{and} \quad \left\langle \tilde{\xi}_k \tilde{\xi}_l \right\rangle = T\, t_s\, \delta_{kl} \,. \tag{3.4}$$

Applying the discrete Fourier transform to the discrete equation of motion of the bead Eq. (3.2) and calculating the expected value of the power spectrum P_f leads to the discrete aliased Lorentzian [4]

$$P_f^x = \frac{2(\Delta x)^2 t_s}{1 + c^2 - 2c \cos 2\pi f t_s} \,, \tag{3.5}$$

where f is the discrete non-negative frequency (in units of Hz). In the continuous-time limit $t_s \rightarrow 0$, the discrete aliased Lorentzian converges to the usual Lorentzian [5], as shown below.

A Taylor expansion of c and $\cos 2\pi f t_s$ for small t_s gives,

$$c^2 = 1 - 4\pi f_c t_s + 4\pi^2 f_c^2 t_s^2 + \cdots \tag{3.6a}$$

$$\cos 2\pi f t_s = 1 - 2\pi^2 f^2 t_s^2 + \cdots \,. \tag{3.6b}$$

Then the discrete-time Lorentzian becomes

$$P_f^x = \frac{\left(4\pi f_c t_s + 4\pi^2 f_c^2 t_s^2\right) \times (D t_s / 2\pi f_c)}{2 - 4\pi f_c t_s + 4\pi^2 f^2 t_s^2 - 2(1 - 2\pi f_c t_s)(1 - 2\pi^2 f^2 t_s^2)} \,. \tag{3.7}$$

Simplifying the terms in the numerator and denominator, we find

$$P_f^x = \frac{2 D t_s^2 (1 - t_s)}{4\pi t_s^2 (f_c^2 + f^2 + f_c f^2 t_s)} \,, \tag{3.8}$$

which, in the continuous-time limit $t_s \rightarrow 0$, converges to the Lorentzian, given by

$$P^x(f) = \frac{D}{2\pi^2 \left(f_c^2 + f^2\right)} \,. \tag{3.9}$$

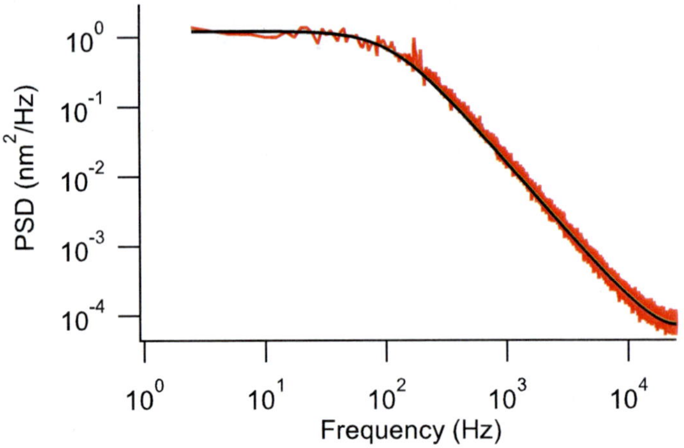

Fig. 3.8 Power spectrum of the displacement of a 3-μm bead from its mean position, in a static trap. The experimental data (red) is fit to Eq. (3.11) (black). The fit parameters are $f_c = 113 \pm 1$ Hz, $D = 0.167 \pm 0.001\ \mu\text{m}^2$/s, and measurement noise $\sigma_m = 0.54 \pm 0.08$ nm

Now, we measure the power spectrum of the noisy measurements P_f^y. For noisy measurements,

$$y = x + \nu\,, \tag{3.10}$$

where the measurement noise ν is Gaussian white noise. The power spectrum of the noisy position y is

$$P_f^y = P_f^x + P^\nu\,, \tag{3.11}$$

where P^ν is the frequency-independent noise floor, and $\sigma_m^2 = P^\nu/2t_s$ the variance of the measurement noise. Equation (3.11) uses the property that the measurement noise is independent of both the bead's position x and thermal noise ξ.

We measure the diffusion constant D, trap stiffness $\kappa = 2\pi f_c k_B T/D$ and the measurement noise σ_m by fitting the power spectrum of the bead in a static trap. Figure 3.8 shows the power spectrum of a trapped 3-μm bead (red) and the fit to the discrete Lorentzian Eq. (3.11) (black).

3.5 Measurement-Noise Control

We control the measurement noise of the bead's position by changing the intensity of the detection laser. The laser intensity is controlled by an acousto-optic modulator (AOM, AA Opto-Electronic, MODA110-B4-40). The AOM is similar to the AOD

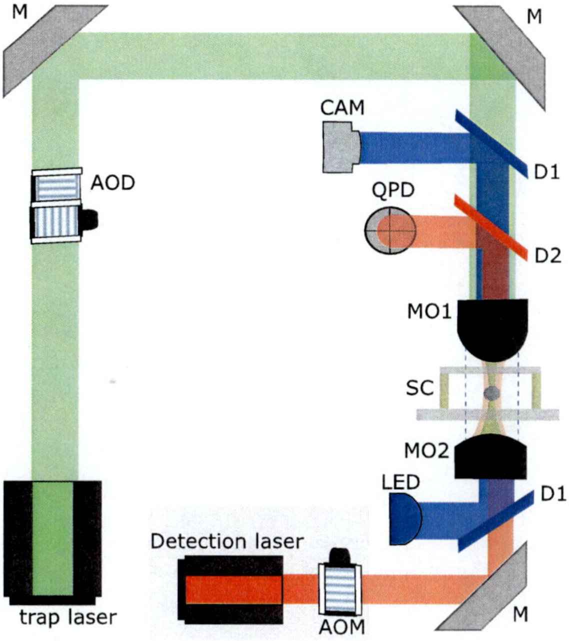

Fig. 3.9 Schematic diagram of the experimental apparatus with measurement noise control (shaded grey region). *M* mirror, *AOD* acousto-optic deflector, *AOM* acousto-optic modulator, *D1* blue dichroic filter, *D2* red dichroic filter, *SC* sample chamber, *MO1* trapping microscope objective, *MO2* detection microscope objective

but controls only the intensity of the laser beam. This is because AOM crystal is connected to a transducer that generates acoustic waves of controllable amplitude, at a single frequency.

The AOM is introduced in the path of the detection laser, as shown in Fig. 3.9. The AOM is aligned such that the first-order Bragg diffraction beam is introduced into the trapping plane. The measurement noise is increased by reducing the intensity of the laser beam. For a good control of the measurement noise and to realize large measurement noise strengths, the detection laser must be carefully aligned. Any stray light from the zero-order diffraction or any other source must be blocked. We thus check that the total intensity on the QPD is zero when the AOM voltage is set to zero. Figure 3.10 shows the power spectral density of the bead in a static trap at different values of the signal-to-noise ratio (SNR), defined as $\sigma/\sigma_{\mathrm{m}}$. The noise floor of the PSD rises as the measurement noise is increased.

3.6 Kalman-Filtered Position Estimate

Experiments are generally limited by the measurement noise in the apparatus, in particular by the noisy measurement of particle position. Measurement noise is usually neglected when it is much smaller than the signal. But when the noise strength is large, a noisy measurement is no longer a good estimate of the system state—in our case, the bead position. A common strategy is to average noisy

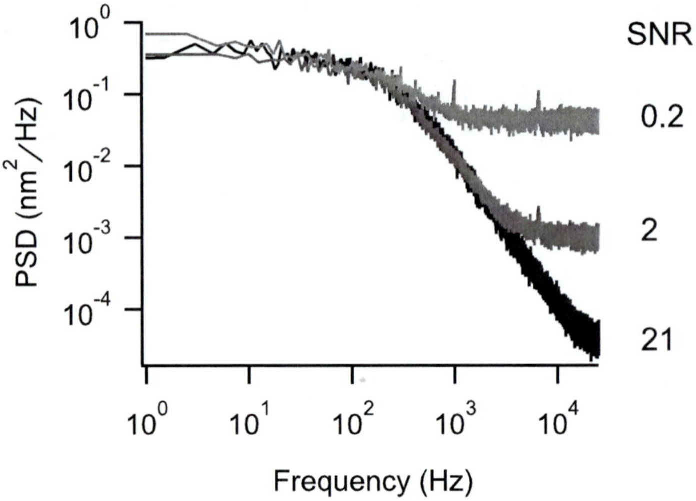

Fig. 3.10 Power spectral density of the measured position of a 4-μm bead, with diffusion constant $D = 0.12\,\mu\text{m}^2/\text{s}$, in a static trap of stiffness $\kappa = 40\,\text{pN}/\mu\text{m}$ at different signal-to-noise ratios (SNR)

measurements to improve the estimates. But, for a bead that undergoes stochastic motion due to thermal fluctuations, a simple averaging strategy will smear features of the bead's actual trajectory. In this section, we discuss a better method of estimating the position of the bead, the Kalman filter, and then benchmark its performance in experiments.

3.6.1 *Kalman Filter Estimator*

For a noisy measurement of a fluctuating bead, the position estimate can be improved by designing an "observer". An observer follows the same equation of motion as the bead, evolving under the same (known) external forces as the bead, but without any noise. The equation of motion of the observer is given by (cf. [6], Chap. 8)

$$\hat{x}_{n+1} = \mathrm{e}^{-t_s}\hat{x}_n + \left(1 - \mathrm{e}^{-t_s}\right)(\lambda_n - \delta_g) + L_{n+1}(y_n - \hat{y}_n)\,, \tag{3.12a}$$

$$\hat{y}_n = \hat{x}_n\,, \tag{3.12b}$$

where $\hat{x}$ is the estimated position, $y\ (= x + \nu)$ the noisy measurement and L_{n+1} the *observer gain.* The gain provides feedback that forces the observer to synchronize with the observed state of the system. Note that we can measure all the experimental parameters required to set up an observer using the power-spectrum analysis, discussed in Sect. 3.4.

The Kalman filter uses the "best" estimate of the bead's position, which is found by tuning the observer gain L to minimize the variance of the error $e_k = x_k - \hat{x}_k$,

$$P_{n+1} \equiv \left\langle e_{n+1}^2 \right\rangle = \left\langle (x_{n+1} - \hat{x}_{n+1})^2 \right\rangle . \tag{3.13}$$

To evaluate the minimum, we start by evaluating the error e_{n+1}

$$\begin{aligned} e_{n+1} &= x_{n+1} - \hat{x}_{n+1} \\ &= \mathrm{e}^{-t_s} e_n + \sqrt{1 - \mathrm{e}^{-2t_s}}\, \xi_n - L_{n+1} e_n - L_{n+1} \nu_n . \end{aligned} \tag{3.14}$$

Using the above equation, we find

$$\begin{aligned} P_{n+1} &= \left\langle (e_{n+1})^2 \right\rangle \\ &= \mathrm{e}^{-2t_s} P_n + (1 - \mathrm{e}^{-2t_s}) + L_{n+1}^2 P_n + L_{n+1}^2 \sigma_m^2 - 2 L_{n+1} P_n \mathrm{e}^{-t_s} . \end{aligned} \tag{3.15}$$

The optimal L_{n+1} that minimizes the variance is found by differentiating P_{n+1} with respect to the gain L_{n+1} and is given by

$$\frac{\mathrm{d}P_{n+1}}{\mathrm{d}L_{n+1}} = 2 L_{n+1} P_n + 2 L_{n+1} \sigma_m^2 - 2 P_n \mathrm{e}^{-t_s} = 0 . \tag{3.16}$$

The optimal observer gain L_{n+1}^*, also called the optimal Kalman gain, is

$$L_{n+1}^* = \frac{P_n}{P_n + \sigma_m^2} \mathrm{e}^{-t_s} . \tag{3.17}$$

If the experimental parameters do not change with time, the variance of the error P_n will tend to a steady-state value P^*. The steady-state value is obtained by eliminating the time dependence and solving the equation

$$\begin{aligned} P^* &= \mathrm{e}^{-2t_s} P^* + (1 - \mathrm{e}^{-2t_s}) + (L^*)^2 P^* + (L^*)^2 \sigma_m^2 - 2 L^* P^* \mathrm{e}^{-t_s} \\ &= \mathrm{e}^{-2t_s} P^* + (1 - \mathrm{e}^{-2t_s}) - \frac{(P^*)^2}{P^* + \sigma_m^2} . \end{aligned} \tag{3.18}$$

The solution to the above equation is

$$P^* = \frac{1}{2}\left((1 - \sigma_m^2)(1 - \mathrm{e}^{-2t_s}) + \sqrt{(1 - \sigma_m^2)^2 (1 - \mathrm{e}^{-2t_s})^2 + 4\sigma_m^2 \left(1 - \mathrm{e}^{-2t_s}\right)} \right) . \tag{3.19}$$

For fast sampling time ($t_s \ll 1$), the first-order approximation of P^* is

$$P^* \approx (1 - \sigma_\mathrm{m}^2)\, t_\mathrm{s} \times \left[1 + \sqrt{1 + \frac{2\sigma_\mathrm{m}^2}{\left(1 - \sigma_\mathrm{m}^2\right)^2 t_\mathrm{s}}}\,\right] . \tag{3.20}$$

The corresponding optimal steady-state observer gain is

$$L^* = \frac{P^*}{P^* + \sigma_m^2} \mathrm{e}^{-t_\mathrm{s}} . \tag{3.21}$$

In practice, we always use the steady-state observer gain L^*, discarding any initial data where the observer-system dynamics are not in steady state.

3.6.2 *Kalman Filter Performance*

The filtered position of the bead is best estimated from noisy measurements by using the Kalman filter. For our system, with Gaussian probability density function and linear forces acting on the bead, the optimal Bayesian estimate is the Kalman filter [6]. For our feedback structure, with feedback delay, the optimal position estimate is given by the predictive Kalman filter, which predicts the current position of the bead given a measurement history that ends one time step before. The estimated position is

$$\hat{x}_{n+1} = \mathrm{e}^{-t_\mathrm{s}} \hat{x}_n + (1 - \mathrm{e}^{-t_\mathrm{s}})(\lambda_n - \delta_\mathrm{g}) + L^*(y_n - \hat{x}_n) , \tag{3.22}$$

where $L^* \in (0, 1)$ is the steady-state filter gain, which depends on the sampling time t_s and signal-to-noise ratio $\mathrm{SNR} \equiv 1/\sigma_m$ (see Sect. 3.6.1). In experiments, we assume that the experimental parameters do not change with time. As a result, L and P in the time-dependent Eqs. (3.15) quickly approach steady-state values. The filter gain indicates how much the measurements can be trusted. For an extremely noisy measurement (SNR$\rightarrow$ 0), the filter gain $L^* \rightarrow 0$. In this limit, the measurements cannot be trusted, and the filter trusts its predictions.

Experiments on the Bayesian information engine were performed using the Kalman-filtered position estimate obtained from Eq. (3.22). Figure 3.11 shows the performance of the experimental implementation of the Kalman filter for a bead in a static trap. The experiment was performed at $\mathrm{SNR} \approx 2$, where the experimental parameters were estimated by fitting the power spectral density using Eq. (3.11). Figure 3.11a shows the trajectory of the estimated bead position (blue), inferred using Eq. (3.22) from the noisy measurements (red). Note that the Kalman filter predicts the position of the bead one time-step ahead, which will be used to apply feedback on the bead in Chap. 6.

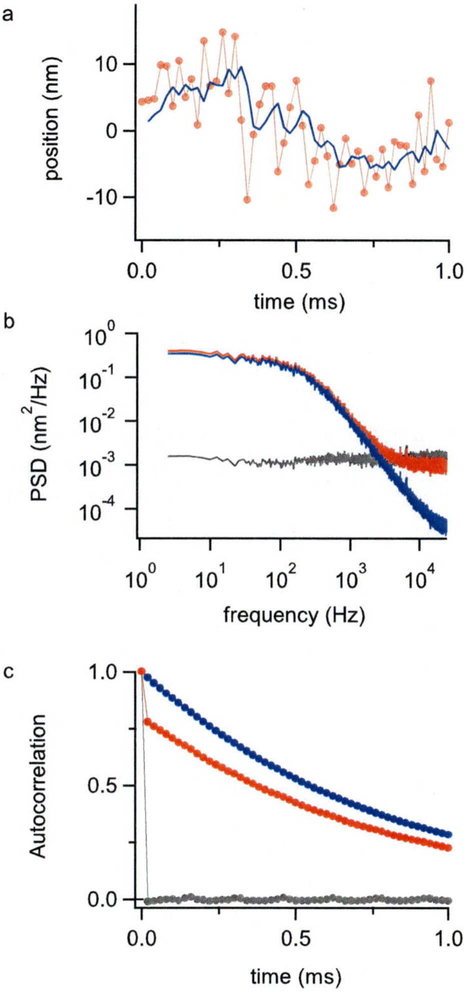

Fig. 3.11 Experimental performance of the Bayesian-filtered estimate. (**a**) Noisy experimental measurement (y, red) and the corresponding Bayesian estimate using Eq. (3.22) ($\hat{x}$, blue) of the bead's position. (**b**) Power spectral density and (**c**) autocorrelation of the noisy measurement (red), Bayesian estimate (blue), and innovations $\hat{x} - y$ (grey). The experiments were performed for SNR ≈ 2, with equilibrium standard deviation $\sigma = 10.1$ nm and measurement noise $\sigma_{\mathrm{m}} = 4.7$ nm

Figure 3.11b shows the power spectral densities of the noisy measurements (red) and Bayesian estimates (blue). The power spectral density (PSD) of the noisy measurements saturates at a finite noise floor (PSD $= 10^{-3}$ nm^2/Hz). The filter-estimated position (blue) has a noise floor less then a tenth that of the individual noisy measurements (red). We demonstrate the quality of the filtered estimate by measuring the innovations, defined as $\varepsilon_n = \hat{x}_n - y_n$, the difference between the estimated position $\hat{x}_n$ and the corresponding measured position y_n. Note that similar expressions such as $\hat{x}_n - x_n$ and $x_n - y_n$ can be computed only from simulations but not from experimental data, because the true position x_n is unknown. The power spectral density of ε_n is flat, showing that the innovations follow a white-noise distribution, and hence all the position correlations arising from the bead's relaxation in the harmonic trap are extracted by the estimator.

Figure 3.11c shows the normalized autocorrelation of the noisy measurements (red), Bayesian estimates (blue), and innovations (grey). Using $y = x + \nu$, the autocorrelation of the noisy measurements is

$$\begin{aligned}\langle y_n y_m \rangle &= \langle (x_n + \nu_n)(x_m + \nu_m) \rangle \\ &= \langle x_n x_m \rangle + \langle x_n \nu_m \rangle + \langle x_m \nu_n \rangle + \langle \nu_m \nu_n \rangle \\ &= \mathrm{e}^{-|m-n|t_\mathrm{s}} + \sigma_\mathrm{m}^2 \delta_{nm} ,\end{aligned} \tag{3.23}$$

where we have used the property that the measurement noise is independent of the true bead position $\langle x_n \nu_m \rangle = \langle x_m \nu_n \rangle = 0$.

The noisy measurements (red curve) show a delta-function peak at $t = 0$ on top of a broader exponential decay, reflecting two processes: First, there is the contribution of measurement noise, which is delta correlated; second, there is the contribution of the dynamics, which has an exponential decay of correlations.

By contrast, the Bayesian estimate (blue curve) has only an exponential decay due to the dynamics—as would the autocorrelation function of noise-free dynamics. In other words, the Bayesian estimator distills the dynamical information in the correlated signal. The innovations (grey curve) are almost entirely delta correlated. Indeed, the degree to which the innovations approximate a delta function serves as a measure of the quality of the Bayesian filter. For example, deviations between a parameter value in the physical system and the corresponding model used for the Bayesian filter produce dynamical correlations in the innovations autocorrelation function [6]. This mismatch has even been used as a way to empirically tune parameters for Bayesian (Kalman) filters ("innovation whitening") [7].

3.7 Dual-Trap Apparatus

We modify the optical feedback tweezers to add a static second trap in the trapping plane for dual-trap experiments. Figure 3.12 shows a simplified schematic diagram of the apparatus. The static trap is created by splitting the green laser beam using a polarizing beam splitter (PBS) and reintroducing it in the feedback-trap's path. An AOD is introduced in one of the laser paths to operate it as a feedback trap. The ratio of laser intensity in each trap is controlled using a $\lambda/2$ waveplate. Both the laser beams are focused in the trapping plane using a single trapping objective (MO1).

We can individually control the separation between the traps and the trap stiffness of each trap. *Trap separation*: The coarse-grained separation of the static trap, with respect to the feedback trap, in the trapping plane is controlled by shifting the laser's path using mirrors; the fine-scale separation is controlled using the AOD. *Trap stiffness*: The stiffness of the static trap is controlled by changing the laser intensity using the $\lambda/2$ waveplate. The stiffness of the feedback trap is controlled by

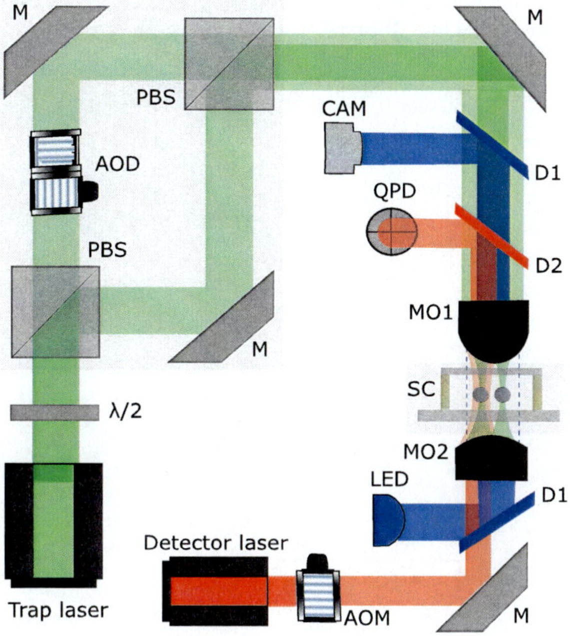

Fig. 3.12 Schematic diagram of the experimental apparatus with dual trap (shaded grey regions). *M* mirror, *PBS* polarizing beam splitter, *AOD* acousto-optic deflector, *D1* blue dichroic filter, *D2* red dichroic filter, *SC* sample chamber, *MO1* trapping microscope objective, *MO2* detection microscope objective

changing the laser intensity using the AOD. Note that a dual trap with different trap stiffnesses can also be realized using time-shared traps [8], which rapidly alternate the centre of the trap between two locations using the AOD. Both the trap position and stiffness can be controlled in the time-shared trap using a well-designed feedback loop. However, to simplify the experimental design, we use separate beams. The challenges of aligning the dual-trap setup are compensated by the simplicity in conceptual design and operation of the experimental apparatus, for the experiments presented in this thesis.

The measurement of the beads' positions is implemented using two separate detectors, operating at different sampling frequencies. We maintain the previously used sampling frequency of 50 kHz ($t_s = 20\,\mu s$) for the feedback trap using the red detection laser and QPD, as shown in Fig. 3.12. This first system can measure the position of only one of the beads (the one held in the left dynamic trap in Fig. 3.12). A second measuring system (camera) can record the position of both beads simultaneously, albeit at lower frequency (1 kHz). The video recorded by the camera is post-processed to extract the bead positions using the circle-fitting LabVIEW algorithm. The image-analysis procedure was automated using LabVIEW. An image of 3-μm beads trapped in the dual tweezers, separated by $\approx 13\,\mu m$, is shown in Fig. 3.13.

A high-frequency position measurement of the bead in the static trap can be achieved using the forward scattered light and a QPD. However, such measurements are inherently prone to cross-coupling from light scattered by the bead in the other

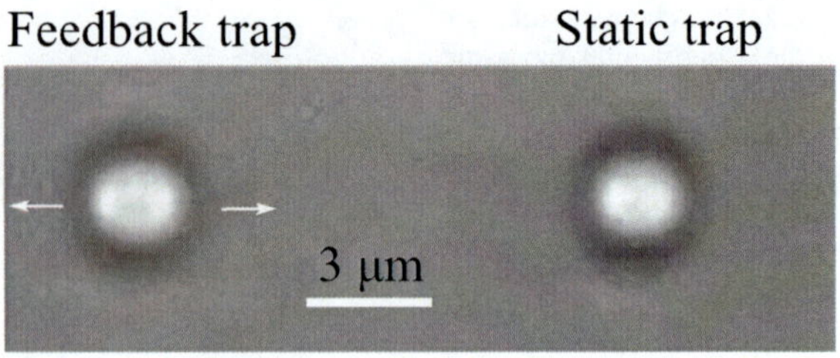

Fig. 3.13 Image of two beads in dual trap. The bead on the left is in the optical feedback tweezers (which moves back and forth, as indicated by the white arrows), and the bead on the right is in a static trap

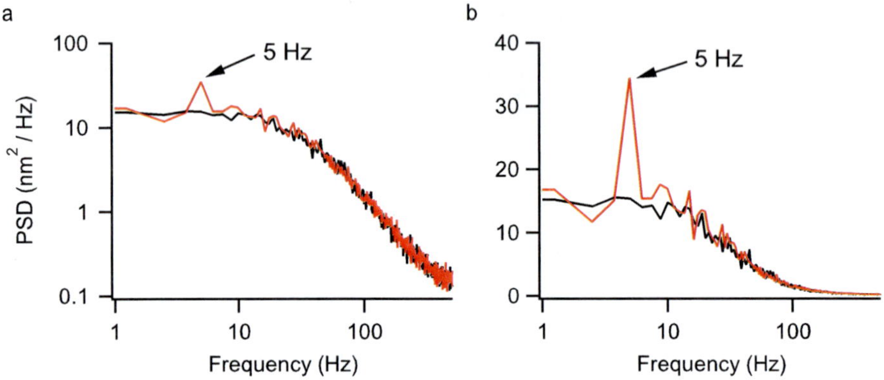

Fig. 3.14 Power spectral density response of the probe bead in a static trap when a driving bead, centred $\approx$ 13 μm away, is in a static (black) and modulating (red) trap. (**a**) and (**b**) represent the experimental data on log-log and semi-log (to emphasize the strength of the 5 Hz peak from hydrodynamic coupling) scales, respectively

trap [9]. Using lasers of different wavelengths for detection removes the cross-coupling but increases the complexity of the apparatus, since there would be three different laser wavelengths (including the trapping laser). The position measurement procedure using a camera is a linear detector (unlike a QPD) and also does not have cross-correlations between the detectors.

In Sect. 2.6, we noted that beads in a dual trap are hydrodynamically coupled. Here, we use the dual-trap apparatus to explore the hydrodynamic coupling of beads. We discuss more-quantitative hydrodynamic coupling experiments in Chap. 8. To demonstrate the hydrodynamic coupling, we modulate the position of the bead in the feedback trap as a sinusoidal wave (without feedback) by changing the position of the trap using the AOD. We record a video of the beads using the camera at a sampling frequency of 1000 Hz. The positions of the beads are extracted using a LabVIEW image-analysis algorithm. The power spectral density of the bead in the static trap (probe bead) shows a peak at 5 Hz (Fig. 3.14), corresponding to the modulation frequency of the bead in feedback trap (driving bead). The peak in the PSD of the probe bead is absent when the driving bead is not modulated. This result shows the presence of hydrodynamic coupling of the bead and that low-frequency hydrodynamic coupling can be measured using our dual-trap apparatus.

3.8 External Electrical Noise Source

In this section, we add an external noise source to the optical trap apparatus. The noise source applies stochastic electric forces on the bead (Fig. 3.15). The sample chamber is modified to add electrodes to apply voltages on the bead. We replace the spacer wires with sticky copper tape (50 μm thick), double-layered to create 100-μm deep sample chambers (Fig. 3.15, inset top-right). The strips of copper tape are long enough to stick out of the glass slide, so that they can be connected to external electrodes using alligator clips. The space between the electrodes is ≈ 5 mm.

To generate the stochastic forces, we use the Johnson noise of a 300-kΩ resistor (TeachSpin, Noise Fundamentals). The TeachSpin noise source has a built-in, two-stage preamplifier whose overall gain is $G_1 = 600$. The noise is further amplified using two other external amplifiers: First, there is a variable-gain amplifier ($G_2 = 10$–10^4) using a TeachSpin noise controller. Second, there is a home-built, fixed-gain ($G_3 = 15$) amplifier that can source the higher current that is needed for this

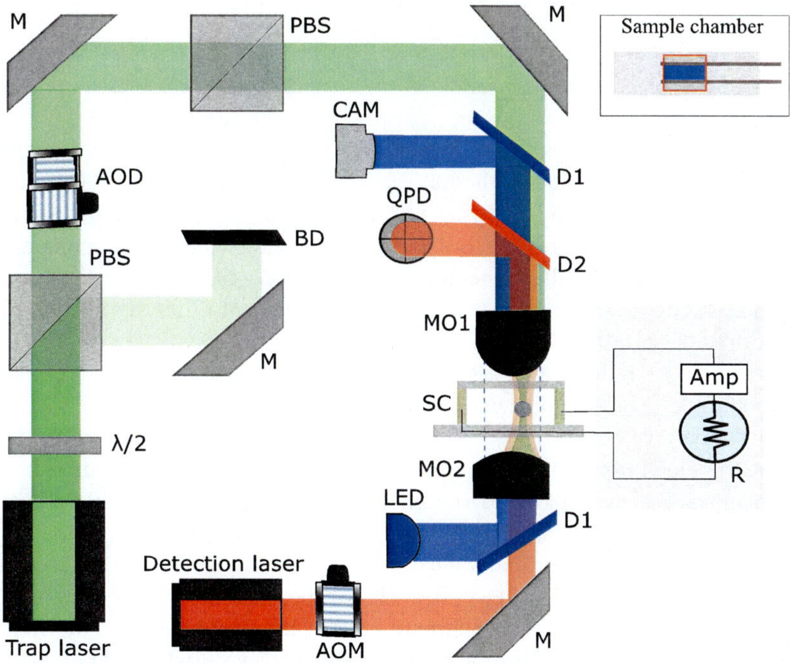

Fig. 3.15 Schematic diagram of the experimental apparatus with external electrical noise (shaded grey region). Inset (right-top): sample chamber with two electrodes. The blue shaded region represents the bead solution. *BD* beam dump, *R* resistor, *Amp* amplifier, *M* mirror, *PBS* polarized beam splitter, *AOD* acousto-optic deflector, *D1* blue dichroic filter, *D2* red dichroic filter, *SC* sample chamber, *MO1* trapping microscope objective, *MO2* detection microscope objective

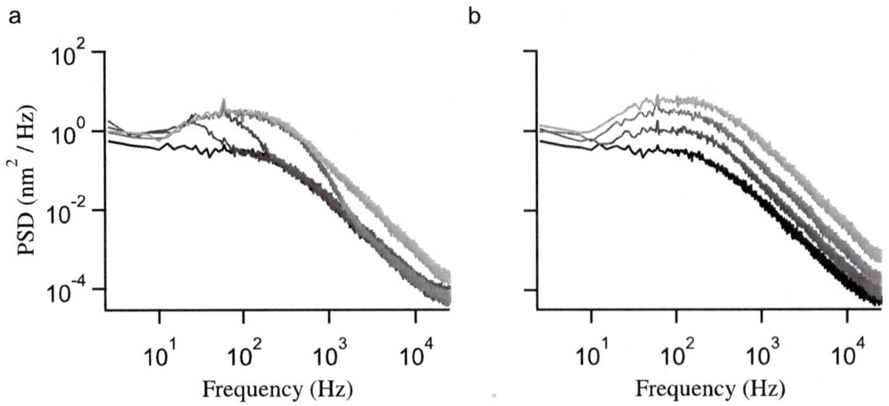

Fig. 3.16 Power spectral density of a trapped bead in the presence of external noise. The black curve denotes the power spectrum in the absence of nonequilibrium noise in both the plots. (**a**) Lighter shades of grey denote a fixed noise amplitude at corner frequencies of the noise: 30 Hz, 100 Hz, 1 kHz, and 100 kHz. (**b**) Lighter shades of grey denote the power spectrum at a fixed corner frequency (100 kHz) with increasing noise amplitude (D_{ne} = 1.6, 6.7, 30.9)

experiment (the two copper electrodes form a capacitor that charges more quickly with higher currents).

We control the bandwidth of the noise using high-pass and low-pass two-pole Butterworth filters included with the TeachSpin controller. The high-pass filter cutoff frequency is set to 10 Hz to prevent electroconvection in the fluid and low-frequency electronic noise. Its bandwidth is sufficiently small compared to the cutoff frequency of the nonequilibrium noise (24 kHz) that calculations of fluctuation strengths that extend to zero frequency do not differ significantly from the experimental results.

The power spectra of trapped beads at different low-pass frequency cutoff and noise amplitude (D_{ne}) are presented in Fig. 3.16a and b, respectively. Experiments were performed for parameter values that did not generate electrochemical breakdown of the electrodes. The measured value of D_{ne} may vary over time because of changes in the electrokinetic mobility of the bead, which would affect the corresponding measured output (presented in Chap. 7). Thus, the output of the information engine, at a particular D_{ne}, is measured immediately after measuring the power spectrum (to infer D_{ne}) [10]. The copper-tape electrodes were used for simplicity; replacing them with platinum electrodes would increase the range of voltages that can be applied to the bead. In Chap. 7, we model the noise source, which is characterized by its corner frequency (f_{ne}) and amplitude (D_{ne}) (see Chap. 7 for the definitions of the parameters). We use the external noise source as a nonequilibrium bath and study the performance of an information engine in a nonequilibrium bath.

In Fig. 3.17, we show the power spectral densities of the background electronic noise (black, using a 1 Ω resistor, whose Johnson noise is negligible relative to the amplifier noise) and 300-kΩ resistor (red, with the background noise). The

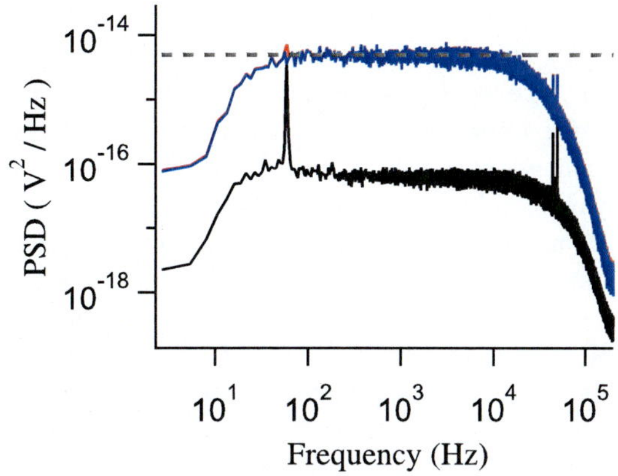

Fig. 3.17 Amplifier noise characterization. Power spectral density of the electronic noise (black), and 300-kΩ resistor with (red) and without (blue) the background electronic noise. Note that the red curve is almost covered by the blue one. Gray dashed line: expected Johnson noise, for $T = 298$ K and $R = 300$ kΩ

high-pass and low-pass filter values were 10 and 100,000 Hz, respectively. The latter low-pass cutoff frequency (100 kHz) is higher than the cutoff frequency of the nonequilibrium noise (24 kHz) and thus does not play a significant role in determining the overall noise properties. The signals were measured using all the amplifiers, with nominal amplifications $G_1 = 600$, $G_2 = 10$, and $G_3 = 15$. The sampling frequency was 400 kHz. The measured output signals were referred back to the respective noise sources by dividing the PSD responses by $(G_1 G_2 G_3)^2$.

The blue curve represents the background-subtracted power spectral density (PSD) of the 300-kΩ resistor. The gray line represents the expected Johnson noise spectral density ($= 4 k_B T R$) for a 300-kΩ resistor, calculated using $R = 300$ kΩ and $T = 298$ K. We find that the experimentally measured response (blue) and expected noise (gray line) are in good agreement. The empirical low- and high-frequency cutoffs are defined as the frequencies where the measured power is reduced by a factor of 1/2. We find 18 Hz and 24 kHz, respectively.

We measure the signal-to-noise ratio (SNR) using the PSDs, where the signal is the Johnson noise of the 300-kΩ resistor and the noise is the background electronic noise. We fit the flat regions of the power spectral densities (black and blue curves) to constants and find from the ratio of the fit values that SNR ≈ 76.8.

3.9 Summary

In this chapter, we discussed the experimental apparatus that is used to perform experiments on information engines in the rest of the thesis. The discussions in this chapter were divided into different sections to discuss the chronological development of the apparatus. The different features were added to perform various studies discussed in Chaps. 4–8. In Chap. 4, we discuss how we use this experimental apparatus to build an information engine that can store gravitational energy and generate directed motion. We use the calibration methods discussed in this chapter to measure the energies exchanged by the information engine. In Chap. 5, we use this apparatus to control the trajectory of a bead using an information engine. Then we use the noise control system to perform studies on an information engine in the presence of measurement noise. The results of this study are discussed in Chap. 6. Next, we use the dual-trap apparatus to measure the hydrodynamic flow generated by a translating bead, as discussed in Chap. 8 and Appendix B. We use the modified sample chamber to apply external electrokinetic forces on the bead and study the performance of an information engine in the presence of external forces, as discussed in Chap. 7. The apparatus has been designed such that any of the experiments presented in this thesis can be performed without modifying the apparatus. This apparatus can be used to study stochastic thermodynamics in the presence of multiple stochastic and deterministic forces. The measurement noise control can be used to study feedback control algorithms in the presence of time-varying measurement and external noise. In the study presented in this thesis, we use electrokinetic forces to generate stochastic forces; however, deterministic feedback forces can also be applied on the bead by connecting the electrodes to the FPGA [11, 12]. Finally, the dual-trap apparatus could also be used to perform experiments to study optimal feedback algorithms to stretch a DNA hairpin [13, 14].

References

1. A. Kumar, Anomalous relaxation in colloidal systems. Ph.D. Thesis, Department of Physics, Simon Fraser University, 2021
2. Y. Jun, J. Bechhoefer, Virtual potentials for feedback traps. Phys. Rev. E **86**, 061106 (6 2012)
3. W.H. Press, S.A. Teukolsky, W.T. Vetterling, B.P. Flannery, *Numerical Recipes: The Art of Scientific Computing*, 3rd edn. (Cambridge University Press, Cambridge, 2007)
4. K. Berg-Sørensen, H. Flyvbjerg, Power spectrum analysis for optical tweezers. Rev. Sci. Instum. **75**, 594–612 (2004)
5. M.C. Wang, G.E. Uhlenbeck, On the theory of the Brownian motion II. Rev. Mod. Phys. **17**, 323 (1945)
6. J. Bechhoefer, *Control Theory for Physicists* (Cambridge University Press, Cambridge, 2021)
7. Q. Wang, W.E. Moerner, An adaptive anti-Brownian electrokinetic trap with realtime information on single-molecule diffusivity and mobility. ACS Nano **5**, 5792–5799 (2011)
8. U. Mirsaidov, et al., Optimal optical trap for bacterial viability. Phys. Rev. E **78**, 021910 (2008)
9. J.-C. Meiners, S.R. Quake, Direct measurement of hydrodynamic cross correlations between two particles in an external potential. Phys. Rev. Lett. **82**, 2211 (1999)
10. T.K. Saha, et al., Information Engine in a Nonequilibrium Bath. Phys. Rev. Lett. **131**, 057101 (2023)
11. A.E. Cohen, W.E. Moerner, Method for trapping and manipulating nanoscale objects in solution. Appl. Phys. Lett. **86**, 093109 (2005)
12. M. Gavrilov, Y. Jun, J. Bechhoefer, Real-time calibration of a feedback trap. Rev. Sci. Instrum. **85**, 095102 (2014)
13. S. Tafoya, S.J. Large, S. Liu, C. Bustamante, D.A. Sivak, Using a system's equilibrium behavior to reduce its energy dissipation in nonequilibrium processes. Proc. Natl. Acad. Sci. U.S.A **116**, 5920–5924 (2019)
14. M. Ribezzi-Crivellari, F. Ritort, Large work extraction and the Landauer limit in a continuous Maxwell demon. Nat. Phys. **15**, 660–664 (2019)

Chapter 4
High-Performance Information Engine

In this chapter, we create and study the performance of a useful information engine that not only extracts energy from heat but also stores energy by raising a weight, as initially imagined by Szilard [1]. The "fuel" for the motor is the information gathered from favourable system fluctuations.

In our study of this information engine, we focus on understanding and then optimizing its performance: How much can it lift? How fast can it go? More precisely, what is the upper bound to the rate of gravitational energy storage and to the directed velocity? We reason that the value of the function of a motor can greatly exceed the cost of running it. For example, in biological applications such as chemotaxis, the metabolic costs of running cellular machinery (including information-processing costs) are usually unimportant compared to the benefit gained by the ability to move toward a new food source or away from a predator [2].

We thus seek to maximize performance, independent of the energy required. As we show, there is a maximum achievable energy-storage rate and a maximum achievable directed velocity, even when the signal-to-noise ratio of the measuring system is arbitrarily high (with correspondingly high costs for information processing); knowing the maximum level of performance independent of information costs can provide a benchmark to evaluate trade-offs between performance and operational costs. We also show that the performance of an information engine is limited by its material parameters. In our case, these parameters include trap stiffness and bead size, and we provide a systematic method for choosing their values to maximize the desired performance measure. Choosing the optimal parameters, we achieve a maximum power of 1000 k_BT/s and maximum velocity of 190 $\mu m/s$. We also find that, by scaling length, time and energy, all the bead diameters follow universal scaling curves for both power and velocity.

Contribution to the chapter: The results presented in this chapter are published in the *Proceedings of the National Academy of Sciences of the United States of America* [3]. This project was done in collaboration with Joseph Lucero, Jannik Ehrich, David Sivak and John Bechhoefer. I developed the information engine

T. K. Saha, *Information-Powered Engines*, Springer Theses,
https://doi.org/10.1007/978-3-031-49121-4_4

experimental apparatus, performed the experiments, performed numerical simulations replicating experimental conditions, developed the data-analysis routine, and analyzed the experimental data. All authors contributed to writing the paper.

4.1 Information Engine Model

Our information engine consists of a 3 μm silica bead, trapped in horizontally aligned optical tweezers, in a water medium (Fig. 4.1b). The water acts as the thermal bath that exchanges heat with the trapped bead. The trap creates a three-dimensional harmonic potential around the bead. The bead's position x fluctuates about an equilibrium position that is below the trap centre λ because of its weight (Fig. 4.1b). Conceptually, the system can be imagined to be a spring-mass system (Fig. 4.1a). The position of the mass is monitored by the "demon". Because of thermal fluctuations, occasionally the bead crosses a predefined threshold, as shown in Fig. 4.1a. After the threshold is crossed, the demon raises the position of the spring, by shifting the black bar. Thus, the fluctuation in the thermal bath (water medium) is converted into stored gravitational energy. Repeating this process, we can store significant amounts of gravitational potential energy by exploiting all the favourable "up" fluctuations.

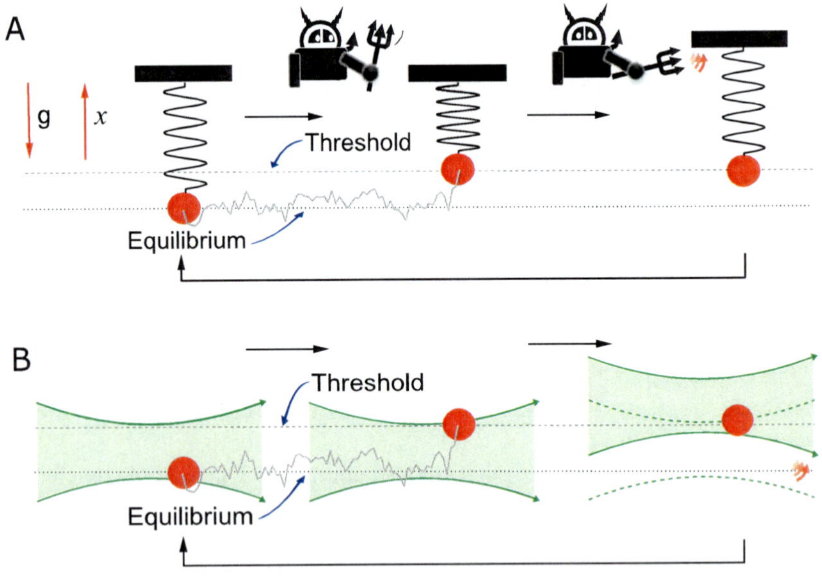

Fig. 4.1 Schematic of the information engine. (**a**) Ratcheted spring-mass system under gravity. (**b**) Experimental realization using horizontal optical tweezers in a vertical gravitational field. Feedback operations on the right side in (a) and (b) are indicated by the small red "swoosh" arrows

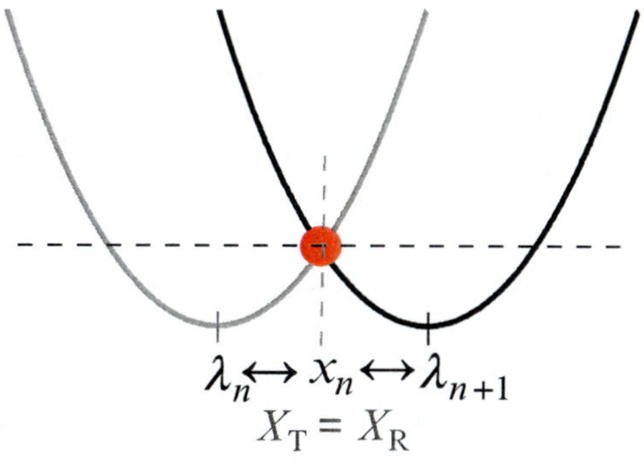

Fig. 4.2 Schematic diagram of realizing zero trap work at ideal feedback gain $\alpha = 2$

The stored gravitational energy can later be used to run a different engine. The engine can be designed such that all the stored energy comes from the thermal fluctuations, without any external work. The external work source is the optical trap. Based on the position to which the trap is raised, in response to the bead fluctuation, the trap can do positive, negative, or zero work on the system (see Sect. 4.3.1).

4.1.1 *Trap Dynamics*

The position of the trap is updated using the feedback rule

$$\lambda_{n+1} = \begin{cases} \lambda_n + \alpha(x_n - \lambda_n), & x_n - \lambda_n \geq X_T \\ \lambda_n, & \text{otherwise}, \end{cases} \tag{4.1}$$

where X_T is the pre-defined threshold, and α is the feedback gain. For a continuous measurement and feedback process, the trap position can be shifted instantaneously such that the bead is reset to an equipotential position to realize zero trap work, as shown in Fig. 4.2. This condition is realized at the ideal feedback gain $\alpha = 2$. However, because of feedback delay and measurement noise, the feedback gain has to be empirically tuned to set the trap power to zero (see Sect. 4.3.1). We study the dependence of α on experimental conditions in more detail in Chap. 6 and find a more robust method for realizing zero trap work.

4.2 Power and Velocity Calculation

We quantify the performance of the information engine by the (long-time average of) directed velocity and stored power, ideally for an infinitely long trajectory. Each trajectory can be viewed as a sequence of independent *ratchet events*, each starting with the particle at position $\lambda - X_R$ inside the trapping potential and ending when the particle fluctuates up and first reaches the position $\lambda + X_{\mathrm{T}}$, where X_{R} is the reset position of the bead after ratchet. The displacement $\Delta x = (\lambda + X_{\mathrm{T}}) - (\lambda - X_{\mathrm{R}}) = X_{\mathrm{R}} + X_{\mathrm{T}}$ is thus fixed for each event, but the time required for event m, the *first-passage time* τ_{FP}, is stochastic [4].

Using the above definitions, we write the velocity

$$v = \lim_{t_{\mathrm{traj}} \to \infty} \frac{X_{\mathrm{traj}}}{t_{\mathrm{traj}}} = \lim_{N_{\mathrm{ratch}} \to \infty} \frac{\sum_{m=1}^{N_{\mathrm{ratch}}} \Delta x}{\sum_{m=1}^{N_{\mathrm{ratch}}} (\tau_{\mathrm{FP}})_m} = \frac{X_{\mathrm{R}} + X_{\mathrm{T}}}{\tau_{\mathrm{MFP}}}, \tag{4.2}$$

where X_{traj} and t_{traj} are the total trajectory length and time and N_{ratch} is the number of ratcheting events. We used the law of large numbers to write $\sum_m (\tau_{\mathrm{FP}})_m \to N_{\mathrm{ratch}} \tau_{\mathrm{MFP}}$, with τ_{MFP} the *mean first-passage time* (MFPT), the mean of τ_{FP}.

The corresponding rate of free energy extraction (power) is $v \, \Delta m \, g$, or, in scaled units,

$$\dot{F} = v \, \delta_{\mathrm{g}} \, . \tag{4.3}$$

For each data point, typically 100 repeated trajectories are measured over a fixed distance of 340 nm. The velocity and power are estimated by replacing X_{traj} and t_{traj} in Eqs. (4.2) and (4.3) with their trajectory averages.

4.2.1 Unbiased Velocity Estimator

We calculate the mean upward velocity of the bead by averaging over N_{traj} trajectories indexed by j. Each trajectory has a different duration; consequently, the number $N(j)$ of position measurements varies for each trajectory. We define the average velocity of ratchet, for varying trajectory lengths, as

$$\langle v \rangle = \frac{\sum_{j=1}^{N_{\mathrm{traj}}} \sum_{n=1}^{N(j)} (x_n - x_{n-1})}{\sum_{j=1}^{N_{\mathrm{traj}}} \sum_{n=1}^{N(j)} t_{\mathrm{s}}}, \tag{4.4}$$

where t_{s} is the sampling time, and x_n denotes the particle position at sampling time t_n.

In the following, we derive Eq. (4.2) from this estimator in the limit of long trajectories. For $N_{\mathrm{traj}} \gg 1$, the law of large numbers permits us to write

$$\langle v \rangle = \frac{N_{\mathrm{traj}} \left\langle \sum_{n=1}^{N} (x_n - x_{n-1}) \right\rangle_{p(N)}}{N_{\mathrm{traj}} \left\langle \sum_{n=1}^{N} t_{\mathrm{s}} \right\rangle_{p(N)}}, \tag{4.5}$$

where $p(\cdot)$ is the probability of the Nth trajectory and we average over the random number N of time samples in a trajectory.

We rewrite this expression in terms of the average over the number N_{ratch} of ratchet events during a trajectory,

$$\langle v \rangle = \frac{\left\langle \sum_{m=1}^{N_{\text{ratch}}} \Delta x_m \right\rangle_{p(N_{\text{ratch}})}}{\left\langle \sum_{m=1}^{N_{\text{ratch}}} \tau_{\text{FP}} \right\rangle_{p(N_{\text{ratch}})}} , \tag{4.6}$$

where Δx_m denotes the bead displacement of ratchet event m, and τ_{FP} is the time interval between successive ratchet events (a first-passage time).

The system quickly reaches a steady state, which implies that all ratchet events are independent and identically distributed events. Steady state is ensured by neglecting the first relaxation time of each trajectory. Then, with $N_{\text{ratch}} \gg 1$, the central limit theorem implies

$$\langle v \rangle = \frac{\langle N_{\text{ratch}} \langle \Delta x \rangle \rangle_{p(N_{\text{ratch}})}}{\langle N_{\text{ratch}} \langle \tau_{\text{FP}} \rangle \rangle_{p(N_{\text{ratch}})}} , \tag{4.7}$$

where the inner averages $\langle \cdot \rangle$ are now taken over a single ratchet event.

Now, $\langle \Delta x \rangle = X_{\text{T}} + X_{\text{R}}$ is an exact relation in the fast-sampling limit, which also implies that the average time to ratchet (i.e., reach X_{T} from $-X_{\text{R}}$) is given by the mean first-passage time: $\langle \tau \rangle \to \tau_{\text{MFP}}$. Thus, the velocity relation in Eq. (4.4) leads to Eq. (4.2),

$$\langle v \rangle = \frac{X_{\text{T}} + X_{\text{R}}}{\tau_{\text{MFP}}} . \tag{4.8}$$

For $X_{\text{T}} = X_{\text{R}}$, the velocity is then given by Saha et al. [3],

$$v(X_{\text{T}}) = \frac{2X_{\text{T}}}{\tau_{\text{MFP}}(X_{\text{T}})} \tag{4.9a}$$

$$\overset{X_{\text{T}} \to 0}{\longrightarrow} \sqrt{\frac{2}{\pi}}\, \mathrm{e}^{-\delta_{\text{g}}^2/2} \left[1 + \operatorname{erf}\left(\frac{\delta_{\text{g}}}{\sqrt{2}} \right) \right]^{-1} . \tag{4.9b}$$

where the second equation is for the limit $X_{\text{T}} \to 0$. The corresponding power is given by Eq. (4.3). In physical units and for large force constants ($\kappa \to \infty$), the velocity and power are [3]

$$v' = \left(\frac{\sigma}{\tau_{\text{r}}} \right) v \overset{\kappa \to \infty}{\sim} \sqrt{\frac{2k_{\text{B}}T}{\pi}} \frac{\sqrt{\kappa}}{\gamma} , \tag{4.10a}$$

$$\dot{F}' = \left(\frac{k_{\mathrm{B}}T}{\tau_{\mathrm{r}}}\right)\dot{F} \overset{\kappa\to\infty}{\sim} \sqrt{\frac{2k_{\mathrm{B}}T}{\pi}}\frac{\sqrt{\kappa}}{\gamma}mg\,. \tag{4.10b}$$

Equations (4.9) and (4.10) were derived by Jannik Ehrich by solving the mean-first-passage time of a bead crossing the threshold X_{T}, starting from $X_{\mathrm{R}} = -X_{\mathrm{T}}$.

4.3 Parameter Optimization

In this section, we sequentially optimize the experimental parameters to maximize the performance of the engine. More precisely, we find the upper bounds on the output power and directed velocity that can be achieved by the information engine.

4.3.1 Trap Work

In our experiments, there is a delay of one time-step between the measurement and feedback update. During this interval, the bead relaxes towards the trap centre. As a consequence, the trap reset needs to compensate for this delay, which would otherwise lead to a significant amount of input trap work.[1] In Fig. 4.3a, we show the dependence of the input trap work for different feedback gain α. In the experiments, α is empirically set to implement the zero-work condition that occurs at the critical feedback gain $\alpha^* \approx 1.5$. Therefore, at $\alpha^* \approx 1.5$, the stored gravitational energy is purely from thermal fluctuations, implementing what we call a *pure information engine*. The trajectory of the bead and trap for the pure information engine is shown in Fig. 4.3b. The position of bead is raised even when the input work is zero, thereby storing gravitational potential energy purely from thermal fluctuations.

4.3.2 Sampling Frequency

To maximize the rate of gravitational-energy extraction (the power) of a pure information engine, we first study its dependence on the sampling frequency. Fixing the trap stiffness κ and hence the relaxation time τ_{r}, we varied the sampling time t'_{s}. Figure 4.4 shows that the power saturates at large sampling frequencies ($f_{\mathrm{s}} = \tau_{\mathrm{r}}/t'_{\mathrm{s}} \gg 1$). Thus, making more measurements per time may not increase the extracted power. Indeed, measurements much faster than the relaxation time τ_{r} of the bead are correlated, and new measurements, on average, fluctuate less and allow

[1] In Chap. 6, we will see that measurement noise also reduces the value of α needed to have zero trap work.

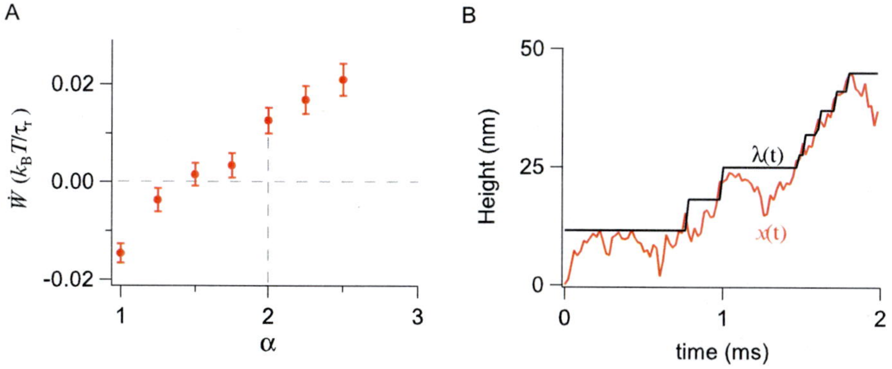

Fig. 4.3 Optimization of trap power. (**a**) Input trap power for different α. The markers denote the experimental values. The error bars denote the standard error of the mean. (**b**) Bead and trap trajectory for $\alpha^* \approx 1.5$ satisfying the condition $\dot{W} = 0$

less energy to be extracted, on average, than do low-frequency measurements [5]. Nonetheless, sampling faster than τ_r reduces the chance of missing a favourable fluctuation.

The solid curve in Fig. 4.4 is based on semi-analytic calculations that use the measured material parameters and agree well with experiments, with no free parameters. Thus, sampling more slowly than the fluctuation time scale τ_r of the dynamics misses possibly useful fluctuations; sampling more quickly eventually yields diminishing returns.

Having established that the extracted power is maximized for infinite sampling frequency, we henceforth use the fastest feedback time of 20 μs, which typically corresponds to scaled sampling frequencies $f_s \gtrsim 100$. Such a scaled sampling frequency is high enough that analytic calculations based on the continuous-sampling limit ($f_s \to \infty$) describe the data well.

4.3.3 Threshold

We next explored how to set the position threshold X_T. This parameter controls the magnitude of the fluctuation that is captured during each ratchet event. The experiments were performed for $\delta_g = 0.8$. The feedback gain $\alpha = 1.9$ ensured that the input power was zero for the chosen threshold values, as confirmed by the grey solid markers in Fig. 4.5.

Figure 4.5 shows that the output power, under the constraint of zero input power, is maximized for $X_T \to 0$ ("continuous ratcheting"). The trap position $\lambda(t)$ then either ratchets to accommodate up fluctuations or pauses when the bead fluctuates down, before reaching the threshold again, as shown in Fig. 4.3b. As X_T increases, the fluctuations that take the bead to the threshold become increasingly

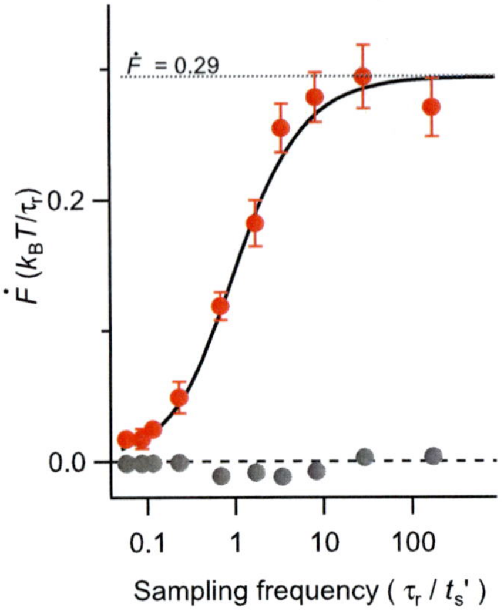

Fig. 4.4 Optimization of sampling frequency f_S. Power as a function of scaled sampling frequency $f_s = \tau_r/t_s'$. Red markers denote experimental values. The black solid curve denotes the semi-analytic results for the same material parameters, feedback gain $\alpha = 1.8$, and threshold $X_T = 0$. For all data, the scaled effective mass $\delta_g = 0.8$, relaxation time τ_r = 3.5 ms, diffusion constant = 0.16 $\mu m^2/s$, trap stiffness κ = 7.3 pN/μm, and bead diameter = 3 μm. The grey markers denote the corresponding $\dot{W}$. Error bars denote the standard error of the mean (Error bars in $\dot{W}$ are smaller than the marker size)

rare (exponentially in X_T), leading to longer wait times between ratchet events; hence, the power tends to zero. The solid black curve is calculated from Eq.(4.9a) by substituting $\tau_{MFP}(X_T)$ [3].

4.3.4 *Trap Stiffness*

Having determined that continuous sampling and continuous ratcheting maximize the extracted power, we explored the role of trap stiffness κ in experiments. We find that power increases monotonically with the trap stiffness, as shown in Fig. 4.6. Note that the output power is in units of k_BT/s. The trap stiffness was increased by increasing the laser intensity. High trap stiffness leads to a shorter relaxation time τ_r of the bead in the trap. The bead relaxes quickly towards the equilibrium position and crosses the threshold sooner. Velocity and power depend inversely on the relaxation time, see Eq. 4.10.

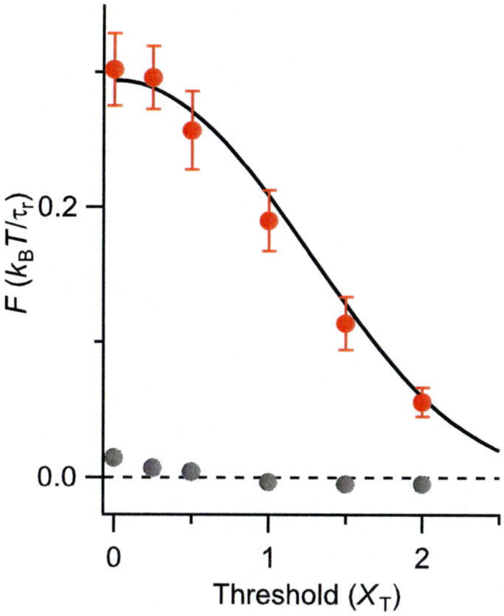

Fig. 4.5 Optimization of ratchet threshold X_T. Power as a function of threshold X_T for fixed $\alpha = 1.9$ and sampling frequency 50 kHz. The grey markers show that the input trap power is small. The black curve follows from Eq. (4.9b). Red markers denote experimental values. For all data, the scaled effective mass $\delta_g = 0.8$, relaxation time $\tau_r = 3.5$ ms, diffusion constant = 0.16 μ m^2/s, trap stiffness $\kappa = 7.3$ pN/ μm, and bead diameter = 3 μm. Error bars denote the standard error of the mean (Error bars in $\dot{W}$ are smaller than the marker size)

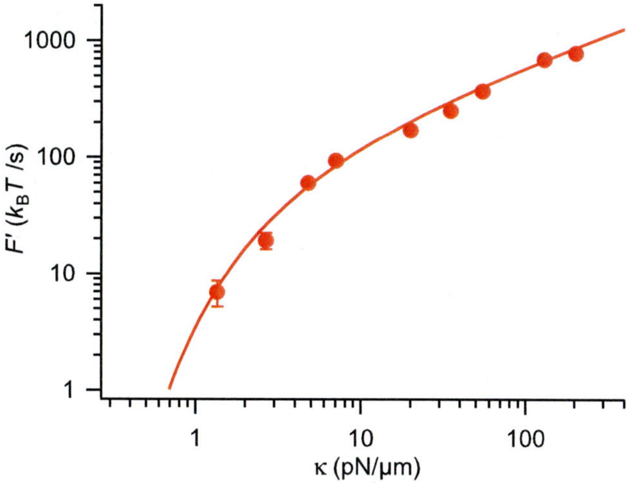

Fig. 4.6 Rate of gravitational free-energy extraction (power) as a function of trap stiffness κ. The markers denote experimental values. The solid line is plotted using Eq. (4.10b). The error bars denote the standard error of the mean

4.3.5 Bead Diameter

Having established the dependence of power on sampling time, threshold, and trap stiffness, we now explore its dependence on bead diameter, using nominal bead diameters of 1.5, 3, and 5 μm. For each trap stiffness κ and for each trapped bead (whose size varies slightly from the nominal size listed by the manufacturer), we determine the value of feedback gain α that makes $\dot{W} \approx 0$. We find that for a fixed

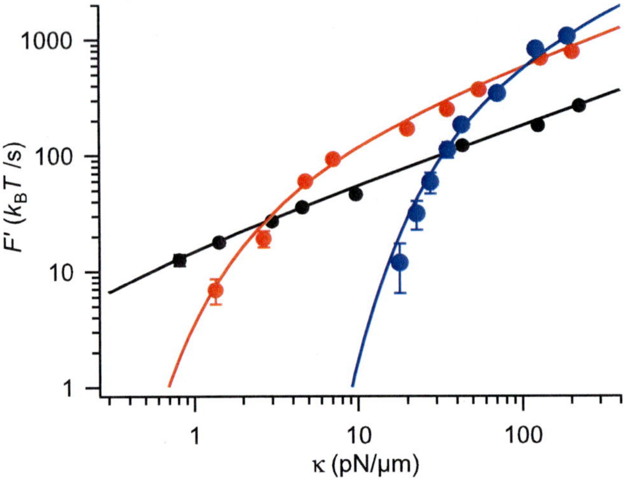

Fig. 4.7 Rate of gravitational free-energy extraction (power) for bead diameters 1.5 (black), 3 (red), and 5 (blue) μm for threshold $X_T = 0$. Markers denote experimental data. The error bars denote the standard error of the mean. The solid curves are plotted using Eq. (4.10b)

trap stiffness, the power is maximized for an optimum bead diameter, as shown in Fig. 4.7. The optimum is the balance of a higher energy gain from each fluctuation of a heavy bead against an increasingly rare fluctuation lifting the bead over the threshold. The maximum trap stiffness is limited by the laser power. In general, a bigger bead diameter maximizes the power at the highest achievable trap stiffness.

4.3.6 Velocity

Here, we measure the directed velocity for the three nominal bead diameters mentioned in the previous section. We find that a smaller bead diameter maximizes the velocity. A smaller bead has a higher diffusion constant, and the information engine rectifies these fluctuations into directed motion. Thus, smaller beads move faster. To maximize the speed, we measure the directed velocity of a 0.5 μm bead and find it produces even higher velocities, for all trap stiffness values, as shown in Fig. 4.8.

For measurements using 0.5 μm beads, we replaced the low-NA condenser lens that focuses the detection beam on the bead with a high-NA water-immersion (NA=1.2) objective. A higher-power red laser (HÜBNER Photonics, Cobolt 06-MLD, 50 mW) was used to increase the scattering signal. The sample chamber was made using two coverslips that were separated by 50 μm spacer wires. The sample chamber was mounted on a 3D printed sample-chamber holder using small magnets.

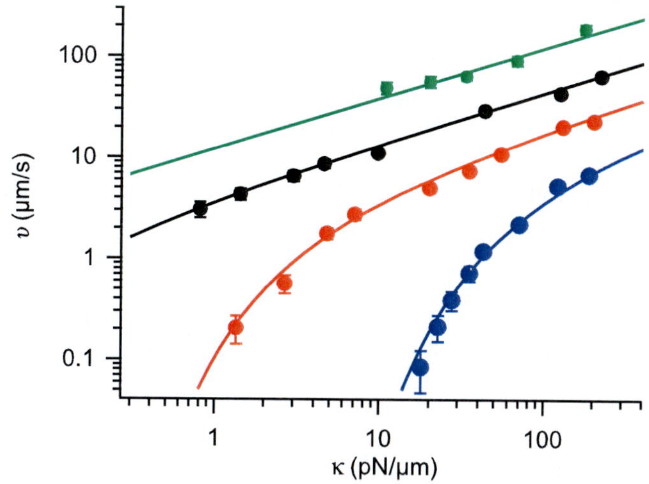

Fig. 4.8 Velocity for bead diameters 0.5 (green), 1.5 (black), 3 (red), and 5 (blue) μ m, for threshold $X_{\mathrm{T}} = 0$, as a function of trap stiffness κ. Markers denote experimental data; solid lines are plots of Eq. (4.10a). Error bars denote the standard error of the mean

Fig. 4.9 Velocities of different beads powered by the information engine compared to other natural and artificial examples of active matter. Inspired by a similar figure in [6]

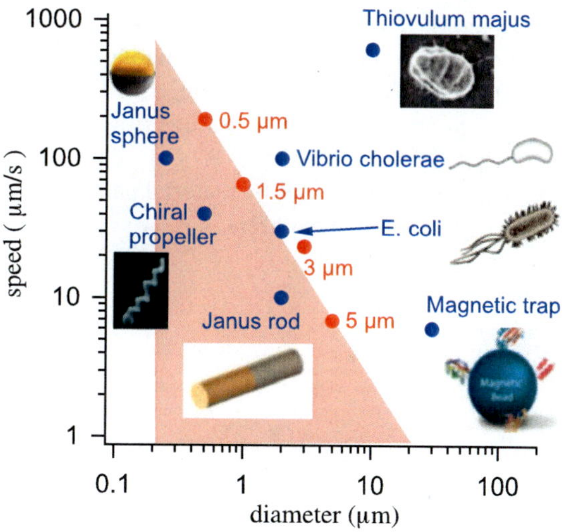

In Fig. 4.9, we compare the velocity achieved by the information engine for different bead diameters to natural and artificial examples of active matter of comparable sizes [6]. The shaded region indicates the achievable velocity range for different bead diameters, extrapolated to smaller and bigger bead diameter range. The velocities achieved by the information engine are comparable to other systems with similar sizes.

4.4 Universal Scaling Law

For fixed trap stiffness and bead density, the velocity decreases with bead size. By contrast, the power is maximized at an optimal intermediate bead size. Scaling the length, time, and energy by the trap standard deviation σ, trap relaxation time τ_r, and $k_B T$, respectively, collapses the data onto single scaled power and velocity curves as a function of the scaled effective mass δ_g, Eq. (4.9b) (Fig. 4.10a, b). The power is maximized at $\dot{F}^* \approx 0.295\ k_B T/\tau_r$ for $\delta_g \approx 0.845$ and the velocity at $v^* \approx 0.8\ \sigma/\tau_r$ for $\delta_g \to 0$. As we saw before, the maximum in extracted power at finite δ_g results from a competition between two effects: the potential energy of a raised object increases with mass, but so does the time to fluctuate beyond a threshold.

4.5 Conclusion

We have designed a simple information-fueled engine that can convert the heat of a surrounding bath into directed motion and hence store gravitational potential energy. A systematic study of conditions that optimize the performance limits of the engine shows that continuous measurements and continuous ratcheting are best. Fortunately, the analysis of the continuous-feedback limit is simpler than that for the corresponding discrete-time dynamics and can draw on well-known results from the analytic theory of mean first-passage times. From the optimization, we find simple expressions for extracted power and velocity, establishing that the performance limits of the engine are set by material parameters such as the stiffness of the spring created by the optical tweezers.

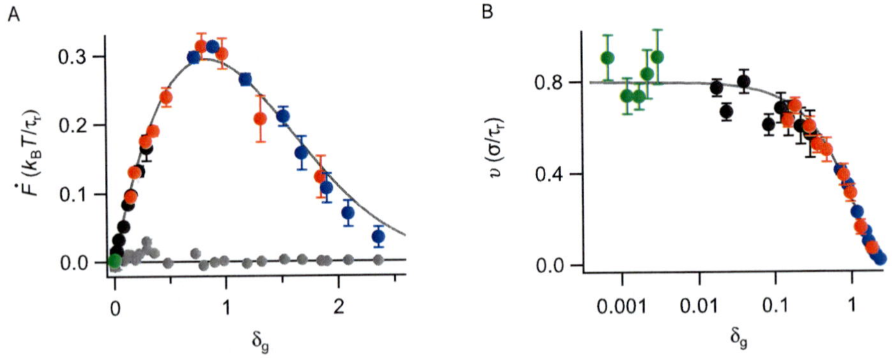

Fig. 4.10 Scaled (**a**) output power and (**b**) velocity for different δ_g. The markers are experimental values for bead diameters 0.5 (green), 1.5 (black), 3 (red), and 5 (blue) μm for threshold $X_T = 0$ and the solid lines are plotted using Eq. (4.9b). The grey markers denote the input trap power. The error bars denote the standard error of the mean

Figures 4.7 and 4.8 show that smaller beads maximize directed motion, but larger beads maximize power extraction. That varying goals call for varying design principles is familiar in macroscopic applications. For example, the diesel engines used in trucks are optimized for power, whereas the turbocharged engines used in race cars are optimized for speed. More generally, systematic connections between material parameters and performance limits are common features of motors. Indeed, motors ranging from proteins to jet engines follow scaling laws whose form is determined by the failure modes of the materials used in the motor construction [7, 8].

By following optimal design principles, we have markedly improved performance relative to previous efforts, which focused instead on information-processing costs and the associated "information-to-work" efficiency of the engine [5, 9–11]. The maximum extracted power is 10^4 times higher than that reported in Ref. [5], although comparable laser powers are used. Most of the improvement in extracted power is achieved through the trap design.

In the present case, power is applied where needed, via a single trap; an array of traps was used in [5]. Our design may also be compared with [10], which uses a single trap, as here, but does not store work. The power levels achieved here exceed those in [12] by an order of magnitude. The improvement relative to [12] arises from careful optimization of parameters (bead size, X_{T}, etc.). Similarly, we increase the directed velocity by a factor of 30 compared to [13] by choosing a smaller bead.

For our setup, the "best" values achieved for power and velocity are 1066 $k_{\mathrm{B}}T$/s and 190 μm/s, respectively. These values are significant: They are roughly ten times faster than *E. coli* and are comparable to the speeds of faster motile bacteria such as those found in marine environments (which need to outswim their algae prey) [14] [Fig. 4.9] and are also comparable to the power used to drive molecular motors such as kinesin [15].

For setups similar to the one used here, the laser power can in principle be increased significantly, which would increase the trap constant κ; however, in many applications, heating will limit the power that can be applied. Another route to increasing performance is to optimize the response properties of the trapped particle. Here, we limited our particle choice to dielectric spheres; more sophisticated core-shell particle designs can reduce beam reflection and scattering forces, thereby increasing the trap stiffness at fixed laser power by a factor of approximately ten [16].

Finally, our information-engine design exploits only the "up" fluctuations. In Szilard's original proposal, the ability to change the connection between mass and partition as a function of the measurement outcome (the side on which the particle is found) allowed exploitation of *all* measurement outcomes. But in our design, "down" fluctuations lead to no feedback response. The information gathered in measuring those fluctuations cannot be exploited, reflecting a structural limitation of the engine [17]. A design that could convert and store energy from all measurements would further enhance information-engine performance.

References

1. L. Szilard, in *Maxwell's Demon 2*, ed. by H.S. Leff, A.F. Rex (IOP Publishing, Bristol, 2003)
2. H.C.E. Berg, Coli, in *Motion* (Springer, New York, 2004)
3. T.K. Saha, J.N.E. Lucero, J. Ehrich, D.A. Sivak, J. Bechhoefer, Maximizing power and velocity of an information engine. Proc. Natl. Acad. Sci. U.S.A **118**, e2023356118 (2021)
4. P. Hänggi, P. Talkner, Reaction-rate theory: fifty years after Kramers. Rev. Mod. Phys. **62**, 251–341 (1990)
5. T. Admon, S. Rahav, Y. Roichman, Experimental realization of an information machine with tunable temporal correlations. Phys. Rev. Lett. **121**, 180601 (2018)
6. C. Bechinger, et al. Active particles in complex and crowded environments. Rev. Mod. Phys. **88**, 045006 (2016)
7. J.H. Marden, L.R. Allen, Molecules, muscles, and machines: universal performance characteristics of motors. Proc. Natl. Acad. Sci. U.S.A. **99**, 4161–4166 (2002)
8. H. Hess, G. Saper, Engineering with biomolecular motors. Acc. Chem. Res. **51**, 3015–3022 (2018)
9. M. Ribezzi-Crivellari, F. Ritort, Large work extraction and the Landauer limit in a continuous Maxwell demon. Nat. Phys. **15**, 660–664 (2019)
10. G. Paneru, D.Y. Lee, T. Tlusty, H.K. Pak, Lossless Brownian information engine. Phys. Rev. Lett. **120**, 020601 (2018)
11. S. Toyabe, T. Sagawa, M. Ueda, E. Muneyuki, M. Sano, Experimental demonstration of information-to-energy conversion and validation of the generalized Jarzynski equality. Nat. Phys. **6**, 988–992 (2010)
12. G. Paneru, et al., Optimal tuning of a Brownian information engine operating in a nonequilibrium steady state. Phys. Rev. E **98**, 052119 (2018)
13. D.Y. Lee, J. Um, G. Paneru, H.K. Pak, An experimentally-achieved informationdriven Brownian motor shows maximum power at the relaxation time. Sci. Rep. **8**, 1–6 (2018)
14. G.M. Barbara, J.G. Mitchell, Bacterial tracking of motile algae. FEMS Microbiol. Ecol. **44**, 79–87 (2003)
15. T. Ariga, M. Tomishige, D. Mizuno, Nonequilibrium energetics of molecular motor kinesin. Phys. Rev. Lett. **121**, 218101 (2018)
16. A. Jannasch, A.F. Demirörs, P.D.J. van Oostrum, A. van Blaaderen, E. Schäffer, Nanonewton optical force trap employing anti-reflection coated, high-refractiveindex titania microspheres. Nat. Photon. **6**, 469–473 (2012)
17. S. Still, Thermodynamic cost and benefit of memory. Phys. Rev. Lett. **124**, 050601 (2020)

Chapter 5
Trajectory Control Using an Information Engine

In Chap. 4, we presented an information engine that could raise an optically trapped bead against gravity and store gravitational potential energy. We found that the average velocity of the bead depends on experimental parameters such as trap stiffness and bead diameter. Here we operate the information engine to convert thermal fluctuations into directed motion, without storing any gravitational energy. We term such an engine a *transporter engine*. Although the ensemble moves at an average velocity, individual trajectories have velocities that fluctuate independently about the average. As a consequence, the algorithm described in Chap. 4 cannot control individual trajectory paths. In this chapter, I explore a feedback algorithm that can make the bead trajectory track a desired path, with zero input trap work. I characterize the performance of the feedback algorithm, drawing on ideas from control theory. I find that the frequency range over which the heuristic feedback algorithm can track is comparable to the corner frequency of the trap.

Contribution to the chapter: The results presented in this chapter are published in the *Proceedings of the SPIE* [1]. John Bechhoefer posed the challenge of controlling bead trajectories without doing any work on the system. I developed the feedback algorithm, performed the experiments, and analyzed the data. We wrote the paper together.

5.1 Horizontal Velocity

As we are interested in controlling the trajectory of the bead as a function of time, we will explore bead manipulation in the horizontal direction, perpendicular to gravity. Eliminating the effects of gravity simplifies the equation of motion of the bead, although the same feedback algorithm can be used to raise the bead, against gravity. A schematic of the horizontal microscopic spring-mass system in a thermal bath is shown in Fig. 5.1a. Because of thermal noise, the mass fluctuates about an

T. K. Saha, *Information-Powered Engines*, Springer Theses,
https://doi.org/10.1007/978-3-031-49121-4_5

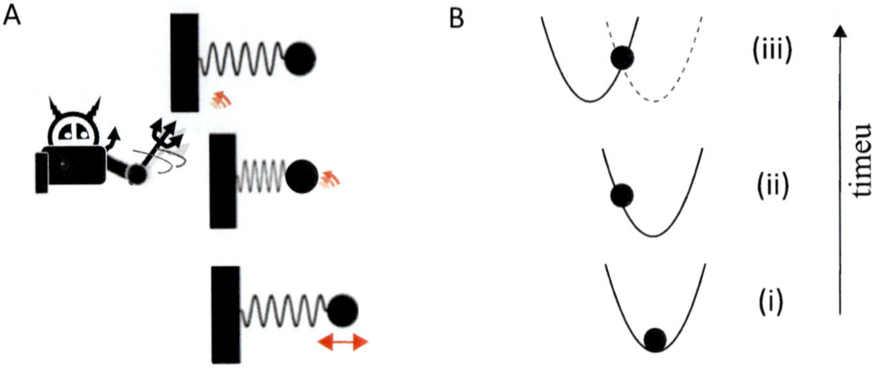

Fig. 5.1 Schematic diagram of the information engine concept. (**a**) Ratcheted spring-mass system moves to the left; (**b**) equivalent dynamics in a harmonic well

equilibrium position. Let the desired directed velocity be to the left. Occasionally, the bead fluctuates left, compressing the spring. The favourable "left" fluctuation is captured by immediately shifting the trap left, decompressing the spring. The corresponding energy of the spring-mass system is shown in Fig. 5.1b. The energy is visualized as a bead in a harmonic well. As with the previously discussed vertical information engine, the distance the spring, i.e. the bottom of the harmonic well, is shifted should be chosen so that no external work is done by the spring on the bead. This condition is ensured by shifting the minimum of the well so that the bead is reset to an equipotential state on the opposite side of the well Fig. 5.1b (iii). In effect, the information engine converts thermal heat into directed motion.

5.2 Bead Dynamics

The equation of motion of the trapped bead in one dimension, moving perpendicular to the direction of gravity, is (cf. Eq. (2.10))

$$\gamma \dot{x}(t) = \underbrace{-\kappa\,(x(t) - \lambda(t))}_{\text{restoring force}} + \underbrace{\sqrt{2k_{\mathrm{B}}T\gamma}\;\xi(t)}_{\text{thermal noise}}\,, \tag{5.1}$$

where $x(t)$ is the position of the bead at time t, $\lambda(t)$ the trap centre, γ the friction coefficient, and $\xi(t)$ is Gaussian white noise, with $\langle \xi(t) \rangle = 0$ and $\langle \xi(t)\xi(t') \rangle = \delta(t - t')$. The corresponding discrete equation of motion of the bead is

$$x_{n+1} = \mathrm{e}^{-t_{\mathrm{s}}/\tau_{\mathrm{r}}} x_n + (1 - \mathrm{e}^{-t_{\mathrm{s}}/\tau_{\mathrm{r}}})\lambda_n + \sqrt{1 - \mathrm{e}^{-2t_{\mathrm{s}}/\tau_{\mathrm{r}}}}\,\xi_n, \tag{5.2}$$

where x_n is the bead's position at the nth time step, $\tau_\text{r} = \gamma/\kappa$ is the trap relaxation time, and ξ_n is an independent Gaussian random variable with zero mean and unit variance. That is, $\langle \xi_n \rangle = 0$, and $\langle \xi_m \xi_n \rangle = \delta_{mn}$.

5.3 Trap Dynamics

In Chap. 4, we found that the velocity of the bead is maximized for $X_\text{T} = 0$. Therefore, for the rest of the thesis I always use the condition $X_\text{T} = 0$. The feedback algorithm in Eq. (4.1) simplifies to

$$\lambda_{n+1} = \begin{cases} \lambda_n + \alpha(x_n - \lambda_n), & x_n > \lambda_n \\ \lambda_n, & \text{otherwise}, \end{cases} \tag{5.3}$$

where α is the feedback gain to tune the input trap power as described in Chap. 4. The velocity of different bead diameters is measured as a function of trap stiffness, as shown in Fig. 5.2. The corresponding vertical velocity is also plotted for comparison. We find that the horizontal and vertical velocities of 0.5 and 1.5 μm beads are equal. For 3 and 5 μm beads, beads moving up are significantly slowed by gravity, as heavier beads stretch the spring and increase the mean first-passage time of the bead to cross the trap centre.

The average velocity of the bead can be tuned by changing the trap stiffness, but individual trajectories cannot be controlled. Figure 5.3 shows the trajectory of the bead, powered by an information engine, along with the corresponding desired trajectory. The slope of the desired trajectory is equal to the average velocity, for a fixed trap stiffness. For the downward ratchet, the ratchet condition is modified to $\lambda_n > x_n$. The bead does not follow closely enough the desired trajectory. As a consequence, we seek to modify the feedback algorithm in Eq. (5.3) so that the bead follows better the desired trajectory, while still satisfying the zero-input-work condition.

5.4 Tracking Algorithm

As mentioned in Sect. 5.1, the bead is propelled by the information engine by exploiting favourable fluctuations. Since these fluctuations occur randomly in time, beads in different trials follow different paths. Figure 5.4a shows individual bead trajectories (light red lines) when the trap position is ratcheted using Eq. (5.3) [2]. For a fixed bead diameter and trap stiffness, each trajectory has a different velocity. Note that individual trajectories diverge from the average (thick red solid line); the

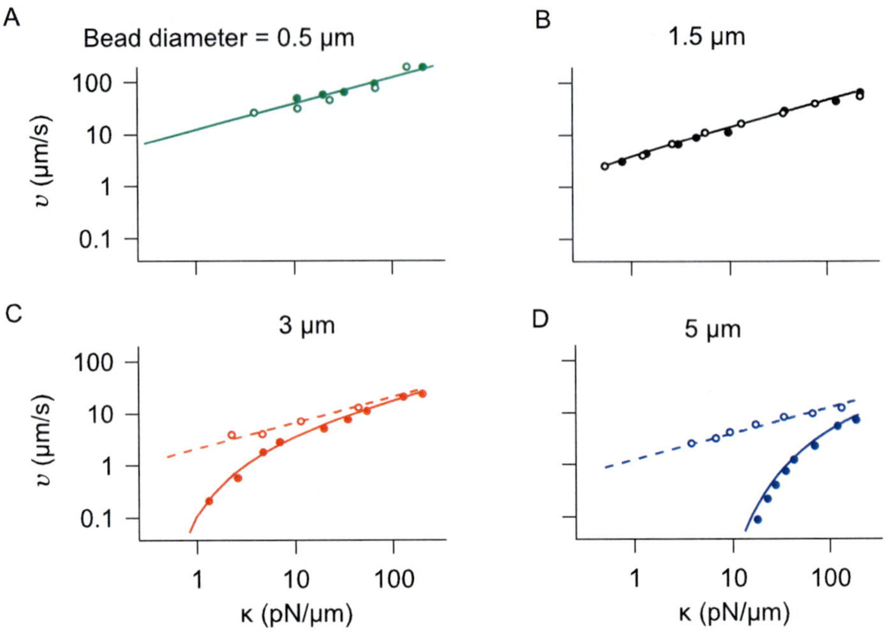

Fig. 5.2 Horizontal velocity (hollow markers) as a function of trap stiffness κ for different bead diameters compared to the corresponding vertical velocity (solid markers). The solid lines for vertical velocities correspond to Eq. (4.10a) and reflect the influence of the gravitational force on upward fluctuations. The dashed lines for horizontal velocities are calculated from Eq. (4.10a), for $\delta_g = 0$

thin red lines represent two standard deviations of the bead trajectories. To restrict the divergence of trajectories with time, I introduce a *tracking feedback algorithm*, given by

$$\lambda_{n+1} = \begin{cases} \lambda_n + \alpha(x_n - \lambda_n)\,, & r_n > x_n > \lambda_n\,,\ \text{or}\ r_n < x_n < \lambda_n \\ \lambda_n\,, & \text{otherwise}\,, \end{cases} \tag{5.4}$$

where α is the feedback gain. Unlike the previous feedback algorithm, Eq. (5.3), the new algorithm, Eq. (5.4), depends on the desired reference position r_n and can track it in either direction. In the previous algorithm, the bead could ratchet only in one direction.

The feedback gain is tuned to $\alpha = 1.7$, which ensures that the average input work is zero (Chap. 4 and [2]). Figure 5.4b shows the average position (solid red line) of the bead and five individual trajectories (light red lines). The trajectories do not diverge with time. The desired velocity in the constrained algorithm was chosen to be slower than the average velocity obtained in the unconstrained case—hence the lower slope in Fig. 5.4b. Figure 5.5 shows that the bead position (red) follows the

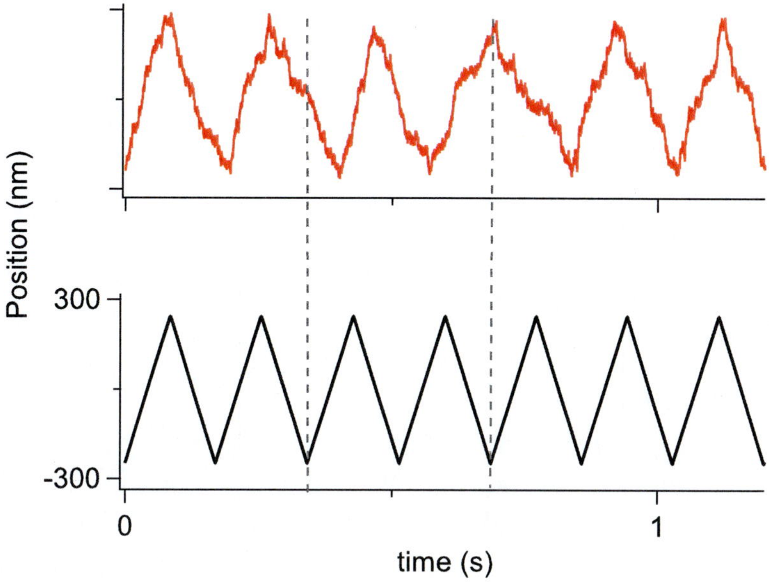

Fig. 5.3 Trajectory of a bead powered (red) by an information engine compared to the desired trajectory (black) using the unconstrained feedback algorithm Eq. (5.3). The dashed lines serve as a guide for the eyes

desired trajectory with the new tracking algorithm. The information engine cannot track velocities faster than the unconstrained feedback algorithm.

A schematic diagram of the feedback algorithm is shown in Fig. 5.6. To understand it, we note that if, after a position measurement, the trap force is already in the "right" direction (i.e., that it will push the particle towards the reference), then no action is taken by the controller. But for forces in the "wrong" direction (away from the reference), the controller ratchets to correct the bead position. Notice that the algorithm ratchets when the bead position is measured to be between the reference and the trap. The algorithm thus captures all fluctuations that push the trap towards the reference signal. Some fluctuations overshoot the reference or push the bead in the opposite direction. In such cases, the trap position stays constant until the next favourable fluctuation. By neglecting the unfavourable fluctuations, the feedback rule constrains the position variance to a narrow band (thin red lines), as shown in Fig. 5.4b. The trap stiffness for all the experiments is 40 pN/ μm, and the corresponding average velocity of the unconstrained information engine is $v_{\max} = 12$ μm/s [2]. I also tried a feedback algorithm that ratcheted after every measurement, so as to always push the particle towards the reference. However, I found the α values required for zero input work led to persistent oscillation of the bead about the desired trajectory.

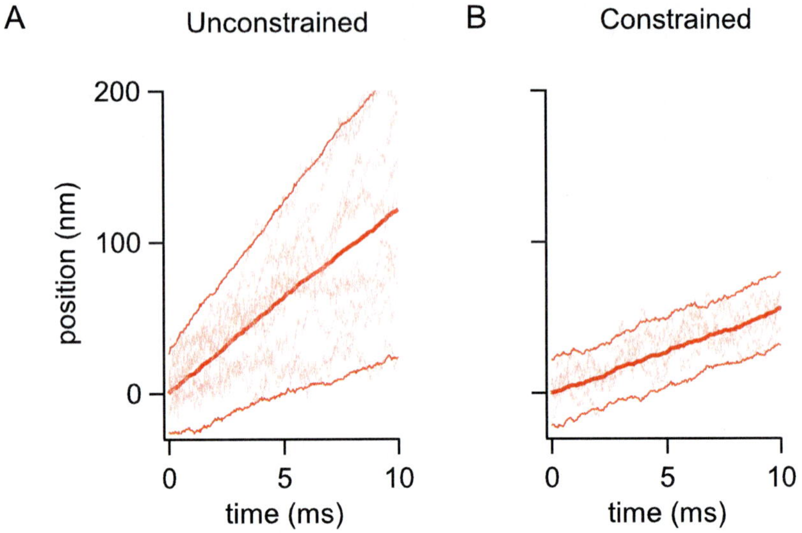

Fig. 5.4 Bead trajectories for constrained and unconstrained feedback algorithms. (**a**) Mean bead position (solid red line shows average over 184 trajectories) for unconstrained feedback algorithm, with ten representative trajectories. (**b**) Mean bead position (solid red line shows average over 132 trajectories) for the conditional bang-bang algorithm, with five representative trajectories. Thin solid red lines represent two standard deviations of the bead trajectories in (**a**), (**b**)

5.5 Trajectory Control

The average velocity of an information engine depends on the trap stiffness and bead diameter [2]. Then a bead powered by an information engine cannot track a reference signal that requires a higher velocity than that allowed by the thermal fluctuations. Figure 5.7 shows the trajectory of the bead (grey) for the desired sine wave reference (black). We can change both the amplitude and the frequency of the sine wave that is followed by the bead. I also show the error signal (residual), $r - x$, for each of the sine waves. For comparison, the position of a bead in a static trap is shown in Fig. 5.7d. This is equivalent to $r(t) = 0$, with $\lambda(t = 0) = 0$, as the trap does not ratchet in this case.

Figures 5.5 and 5.7 show example trajectories where the reference waves are well tracked. In all these cases, the magnitude of the velocity of the reference, $|\dot{r}(t)|$, is always below the maximum unconstrained velocity, $v_{\max}$. However, the algorithm cannot respond to a reference that changes arbitrarily quickly, even when $|\dot{r}(t)| < v_{\max}$.

To quantify the range of frequency the feedback algorithm can track, I measure the transfer function of the controller [3]. The schematic block diagram of the feedback loop is shown in Fig. 5.8. Based on the error $r_n - x_n$, the controller K sends an input to the system G. The controller is the feedback algorithm in Eq. (5.4), and the input is the force due to the trap on the bead. The corresponding output response

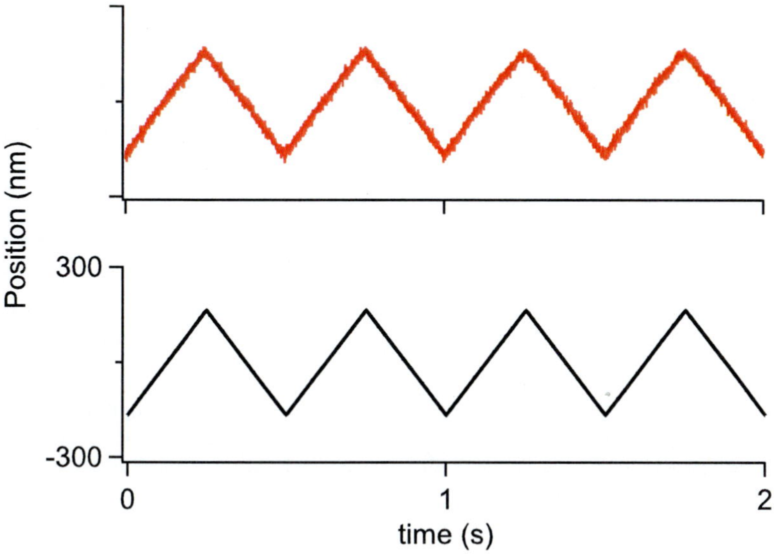

Fig. 5.5 Trajectory of a bead powered (red) by an information engine compared to the desired trajectory (black) for the tracking algorithm in Eq. (5.4)

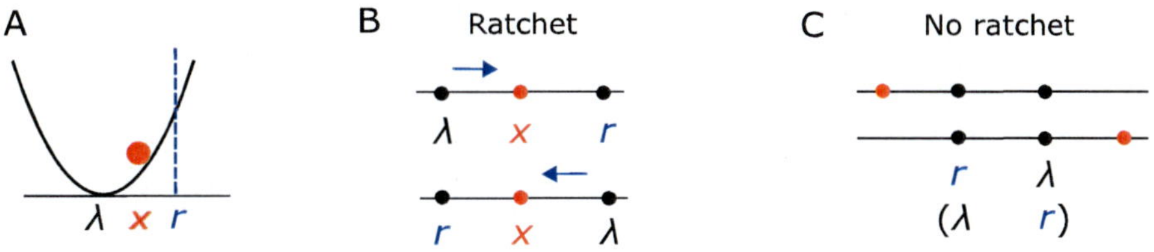

Fig. 5.6 Schematic of the feedback algorithm. (**a**) Trap position λ, bead position x and reference r for a trapped bead. (**b**) Trap position ratchets when $\lambda < x < r$ or when $r < x < \lambda$. (**c**) No ratcheting for other combinations of r and λ, where the red circles represent the bead positions

x, the bead position, is measured. In addition to the reference input, there is also an input to the system that arises from the thermal noise ξ. The closed-loop response of this system is then given by

$$x = \frac{KG}{1+KG} r + \frac{1}{1+KG} \xi \tag{5.5a}$$

$$\equiv T r + S \xi \tag{5.5b}$$

where S and T are the sensitivity and complementary sensitivity functions, respectively [3]. For thermal fluctuations that are smaller than the reference signal, I use the approximation $\frac{\xi(f)}{r(f)} \to 0$, where $r(f)$ and $\xi(f)$ are the Fourier transforms of the

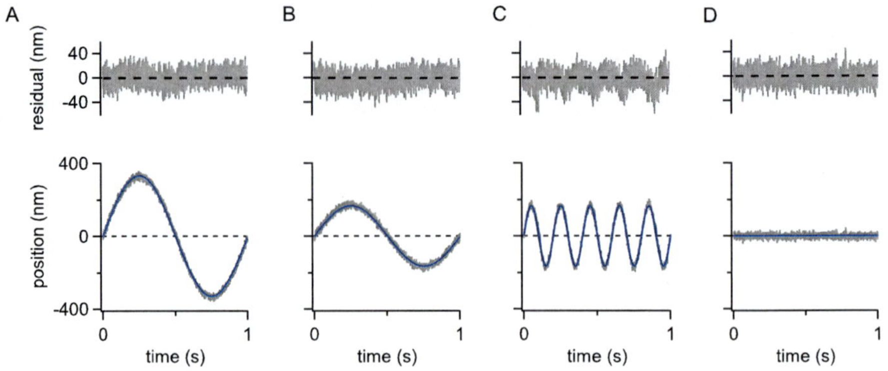

Fig. 5.7 Controlled trajectory of the bead to follow sine waves. The bead position (x, grey) and trap position (λ, blue) for different reference (r, black) signals, and the corresponding residual. The plot represents $r(t) =$ **(a)** $2A\sin(\omega t)$, **(b)** $A\sin(\omega t)$, **(c)** $A\sin(5\omega t)$, and **(d)** $r(t) = \lambda(t) = 0$, where $A = 160$ nm and $\omega = 2\pi \times 1$ Hz

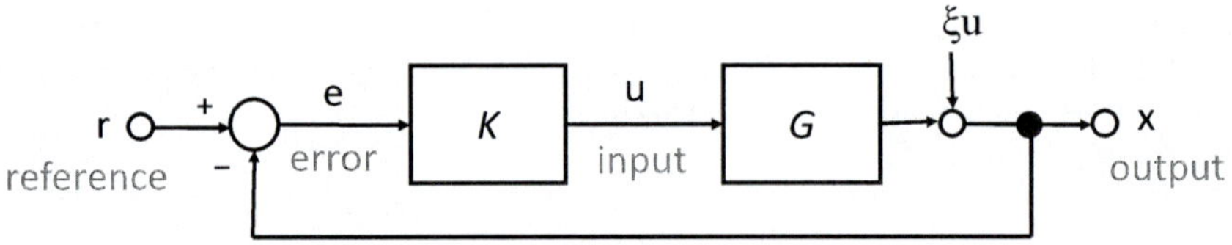

Fig. 5.8 Block diagram of the feedback loop. K is the controller, G the system, and ξ the thermal noise

reference $r(t)$ and thermal noise $\xi(t)$, respectively. The transforms are evaluated at the frequency f. The closed-loop transfer function is then given by

$$T(f) = \frac{x(f)}{r(f)}, \tag{5.6}$$

where $x(f)$ is the Fourier transform of the bead position $x(t)$.[1]

To experimentally measure the transfer function, I input reference sine waves. For each frequency, the amplitude of the reference sine wave is chosen so that it is always below the maximum achievable velocity at the given frequency. In particular, for an input reference $r(t) = A\sin\omega t$, with angular frequency $\omega = 2\pi f$, the amplitude A is chosen to be below $v_{\max}/\omega$. I record the reference signal and the corresponding bead trajectory and then calculate the power spectral density of the

[1] We use an "overloaded" operator notation, where the object x is a time series when we write $x(t)$ and is a Fourier transform when written as $x(f)$. The functional forms are, of course, different in the two cases.

bead position and reference signal. Figure 5.9a shows the power spectral density of the trapped bead, $|x(f)|^2$, when it is driven by thermal fluctuations from the surrounding water bath.

5.6 Measurement of the Feedback Bandwidth

Figure 5.9b shows the reference-bead transfer function, defined as the ratio of the power of the bead to that of the reference, as a function of frequency. We find that the bead follows the reference trajectory at low frequencies but not at high frequencies (blue solid markers). The feedback algorithm thus acts as a low-pass filter. Beyond 1000 Hz, the power of the bead response to the reference is less than the thermal noise and cannot be measured. The measured frequency response is well-fit by a first-order low-pass filter [3]

$$|T|^2 = \left|\frac{x(f)}{r(f)}\right|^2 = \frac{1}{1+(f/f_0)^2}, \tag{5.7}$$

where f_0 is the frequency bandwidth of the algorithm. The DC gain (numerator) is 1, as the reference and bead position are measured in the same units. The fit to

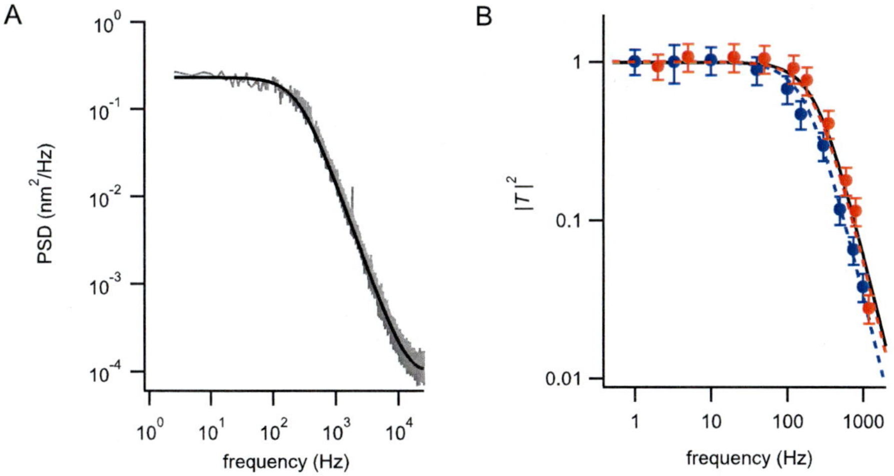

Fig. 5.9 Transfer functions for bead response using the tracking algorithm. (**a**) Thermal noise power spectral density of a trapped bead (grey) calculated from a 20 s time series, divided into 50 0.4-s time series. Black solid line shows fit to an aliased Lorentzian [4] with the corner frequency $f_c = 254 \pm 2$ Hz. (**b**) Experimental data for $|T|^2$ as a function of reference frequency for the information engine (blue solid markers) and open-loop (red solid markers). Blue and red dashed lines are fit to Eq. (5.7), with $f_0 = 186 \pm 11$ Hz for the information engine and $f_0^{\mathrm{OL}} = 240 \pm 13$ Hz for direct driving by sinusoidal reference inputs. Black solid line in (**b**) is the amplitude-normalized fit in (**a**)

Table 5.1 Amplitude and frequency of the reference wave for the information engine (IE) and open-loop (OL) driving

IE	Frequency (Hz)	1	3.3	10	40	100	150	300	500	750	1000
OL	Amplitude (nm)	165	165	165	45.4	17.3	13.2	4.9	3.3	2.0	1.7
	Frequency (Hz)	2	5	20	50	120	180	350	600	800	1200
	Amplitude (nm)	165	165	165	165	165	165	165	82.5	82.5	82.5

the experimental data is shown as the blue dashed line in Fig. 5.9b. For comparison, I measure the transfer function of the trapped bead (red solid markers) by driving the trap with the desired reference sine wave without feedback, i.e., in open-loop. I fit the data to Eq. (5.7), shown as the red dashed line, to find the corresponding bandwidth f_0^{OL}. I find that the bandwidth $f_0 = 186 \pm 11$ Hz of the feedback algorithm is slightly lower than the bandwidth $f_0^{\mathrm{OL}} = 240 \pm 13$ Hz of the open-loop driving. Naively, one might expect that a conventional open-loop driving, without any force constraints, would track the reference significantly better than the force-constrained information engine. However, I find that the open-loop driving has a bandwidth nearly that of an information engine.

The amplitude-normalized power spectrum found from the fit is also shown in Fig. 5.9b (black solid line). Note that the open-loop bandwidth $f_0^{\mathrm{OL}} = 240 \pm 13$ Hz and corner frequency $f_c = 254 \pm 2$ Hz of the trap match well, as they are equivalent methods to measure the bandwidth. Since the bead response, when driven by an information engine, is limited by the response time of the trapped bead to the thermal fluctuations in the bath, a close match between the controller bandwidth $f_0 = 186 \pm 11$ Hz and the corner frequency f_c suggests that any other feedback algorithm will at best only marginally improve the tracking performance. Thus, the information engine's response seems limited by the response time of bead in the trap, and not by the feedback algorithm. The non-normalized amplitude and frequency chosen for the reference sine wave in Fig. 5.9b is given in Table 5.1. As noted earlier, the amplitude of the sine wave for the information engine was chosen to below $v_{\max}/\omega$.

With the feedback algorithm in Eq. (5.4), the bead could follow a sine wave of amplitude 8 nm. The amplitude is comparable to the standard deviation of the bead in the trap, $\sigma = 10$ nm. The bead trajectory (grey) and the reference (black) are shown in Fig. 5.10a. The corresponding power spectrum density, shown in Fig. 5.10b, has a peak at 10 Hz which corresponds to the frequency of the reference sine wave. I found that the bead could also track a sine wave with amplitude 1 nm, an order of magnitude smaller than σ, at 10 Hz (Fig. 5.10c). As the amplitude is smaller than σ, one cannot see in Fig. 5.10c that the bead does in fact follow the reference signal. However, a peak at 10 Hz in the corresponding power spectrum density (Fig. 5.10d), confirms that the bead follows the reference signal. Of course, tracking motion at such a low amplitude is possible only because the desired trajectory is periodic, which allows averaging that reduces the effect of noise. We would expect

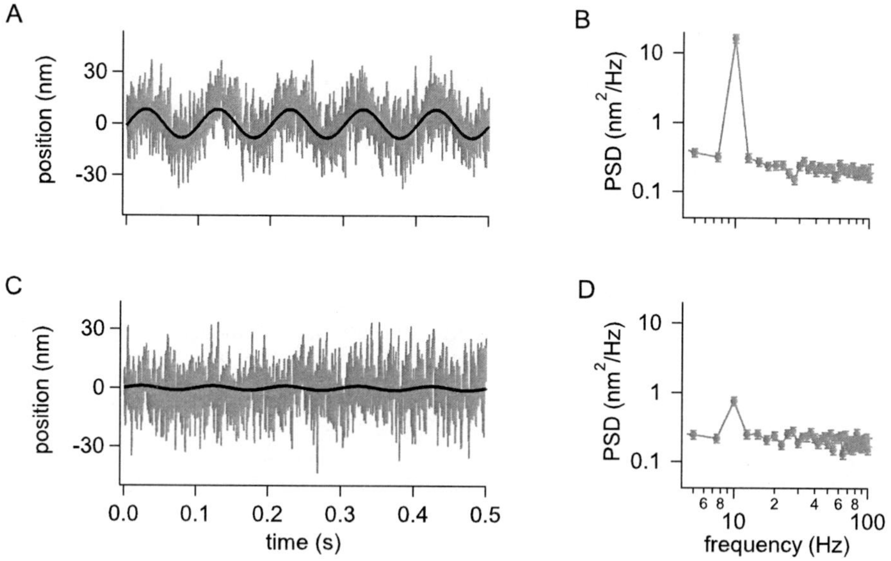

Fig. 5.10 Trajectory control to follow a sine wave. The bead position (grey) for different reference sine waves (black) of amplitude (**a**) 8 nm and (**c**) 1 nm at a frequency of 10 Hz. The standard deviation of the bead in the trap $\sigma = 10$ nm. (**b**), (**d**) shows the corresponding power spectra calculated from a time series of 20 s

to be able to track a non-repeating waveform only if its amplitude exceeds the thermal noise level, σ.

5.7 Conclusion

I have developed a feedback algorithm that can constrain a bead to follow a desired trajectory, with zero input trap work. The feedback algorithm continuously ratchets when the bead lags the reference wave in either direction and stops when it reaches the reference. The algorithm is reminiscent of the bang-bang algorithm from control theory [3] where the controller applies zero or maximum force in either direction to control the dynamics. It might be possible to further improve the tracking bandwidth, but we do not expect to be able to increase the bandwidth much, as our algorithm is already close to the corner frequency f_c of the trap ($f_0/f_c \approx 0.77$). Although conventional feedback algorithms can reduce the position variance of the bead below the trap standard deviation σ, the constraint that the algorithm not allow any work to be done on the particle is a severe one, and it is not clear whether reducing the variance of fluctuations around a reference to values below that of the natural trap is possible for an information-fueled engine.

Controlled manipulation of microscopic systems has potential application in drug delivery [5]. A major challenge for controlling the trajectory of microscopic systems is the thermal fluctuations that push them away from the target. On the contrary, information engines can convert these fluctuations into "useful" motion. With the feedback algorithm presented here, an information engine can control the bead trajectory and limit the position variance to a width comparable to σ. This feature could help transport a bead to a specific location, with a precision of σ, at a time defined to a precision τ_R. For the experiment, recall that $\sigma = 10$ nm and $\tau_r = 0.6$ ms. Experiments have also shown that carefully designed feedback algorithms can reduce the dissipation in stochastic systems [6]. Integrating the ability to efficiently use thermal fluctuations and reducing thermal dissipation could also further improve the performance of presently realized micro-robots [5].

References

1. T.K. Saha, J. Bechhoefer, Trajectory control using an information engine. Opt. Trapp. Opt. Micromanipul. XVIII **11798**, 117980L (2021)
2. T.K. Saha, J.N.E. Lucero, J. Ehrich, D.A. Sivak, J. Bechhoefer, Maximizing power and velocity of an information engine. Proc. Natl. Acad. Sci. U.S.A **118**, e2023356118 (2021)
3. J. Bechhoefer, *Control Theory for Physicists* (Cambridge University Press, Cambridge, 2021)
4. K. Berg-Sørensen, H. Flyvbjerg, Power spectrum analysis for optical tweezers. Rev. Sci. Instum. **75**, 594–612 (2004)
5. J.W. Yoo, D.J. Irvine, D.E. Discher, S. Mitragotri, Bio-inspired, bioengineered and biomimetic drug delivery carriers. Nat. Rev. Drug Discov. **10**, 521–535 (2011)
6. S. Tafoya, S.J. Large, S. Liu, C. Bustamante, D.A. Sivak, Using a system's equilibrium behavior to reduce its energy dissipation in nonequilibrium processes. Proc. Natl. Acad. Sci. U.S.A **116**, 5920–5924 (2019)

Chapter 6
Bayesian Information Engine

In previous chapters, we discussed an information engine that could store energy and generate directed motion [1]. We also presented an algorithm to control the trajectory of the bead without doing any work directly on it [2]. In all the experiments presented so far in this thesis, the feedback was based on high-precision position measurements of the bead. One obstacle that degrades the output of an information engine is inaccurate information about the system that arises from measurement noise. Since information engines respond to measurements of thermal fluctuations, measurement noise can lead to wrong feedback decisions. Feedback actions chosen based on inaccurate measurements reduce the work extracted from the surrounding thermal bath and can even, at high noise levels, lead to a net heating of the thermal bath [3].

Prior efforts to account for noisy measurements in information engines have all used "naive" feedback algorithms based directly on the most recent noisy measurement [3, 4]. Here we show that such information engines, with unidirectional ratchets, have a phase transition between working and non-working regimes: Below a critical level of signal-to-noise ratio for measurements of the engine state, a "pure" information engine—one that requires no work input beyond that needed to run the measurement and control apparatus—is not possible.

Although previous studies noted the degradation of performance due to measurement noise, they did not attempt to alter the feedback algorithm to compensate. Yet theoretical studies have indicated that incorporating the information contained in past measurements via optimal feedback control could greatly improve the performance of an information engine [4–7]. Indeed, experiments in other areas of physics have used feedback that incorporates Bayesian estimators to demonstrate spectacular results, even in the presence of high measurement noise; significant achievements include trapping a single fluorescent dye molecule that is freely diffusing in water [8] and cooling a nanoparticle to the quantum regime of dynamics [9, 10].

T. K. Saha, *Information-Powered Engines*, Springer Theses,
https://doi.org/10.1007/978-3-031-49121-4_6

In this chapter, we discuss an experimental realization of an optimal Bayesian information engine that retains the relevant memory of all past measurements in a single summary statistic. Using the extra information from past measurements and correctly compensating for delays in the feedback loop via predictive estimates, we extract and store significant amounts of energy, even in the presence of high measurement noise.

Our implementation of the Bayesian filter uses the optimal affine feedback control algorithm [11], at optimal experimental parameters [1], to maximize the engine's rate of gravitational-energy storage. The relevant information from past observations is used to minimize the uncertainty in the bead's position. This Bayesian information engine extracts energy even at low signal-to-noise ratio (SNR), avoiding the phase transition in the naive information engine that leads to zero output. Under any conditions, this engine extracts at least as much work as the naive engine and, indeed, reaches the upper bound on the performance of Gaussian information engines.

Contribution to the chapter: The results presented in this chapter have been published in *Physical Review Letters* [12]. The project was done in collaboration with Joseph Lucero, Jannik Ehrich, David Sivak and John Bechhoefer. I performed the experiments, built the measurement-noise control system, developed the data-analysis routine and analyzed experiment data. I found the performance benefit of the information engine using the Bayesian estimate instead of the naive measurements and also discovered that the feedback delay removes bias in work estimate using numerical simulations. Joseph Lucero carried out the numerical simulations presented in this chapter to generate the theoretical predictions; I performed numerical simulations that replicated experimental conditions (not presented in the thesis). John Bechhoefer suggested the presence of a phase transition. Joseph Lucero and I, independently and simultaneously, established the presence of a phase transition in the naive information engine through simulations and experiments, respectively. Jannik Ehrich showed how to calculate analytically the critical value of the signal-to-noise ratio for the phase transition. I led the effort to write the manuscript, with help from all my collaborators.

6.1 Equation of Motion

The optically trapped bead can be modelled by a spring-mass system [Fig. 6.1a]. The true position of the bead is estimated from a noisy measurement y. In the experiments presented in this chapter, measurement noise is increased by reducing the intensity of the detection laser beam (see Sect. 3.5). The discrete-time dynamics of an optically trapped bead is given by (cf. Sect. 2.3)

$$x_{n+1} = \mathrm{e}^{-t_s} x_n + (1 - \mathrm{e}^{-t_s})(\lambda_n - \delta_g) + \sqrt{1 - \mathrm{e}^{-2t_s}}\, \xi_n \,, \tag{6.1a}$$

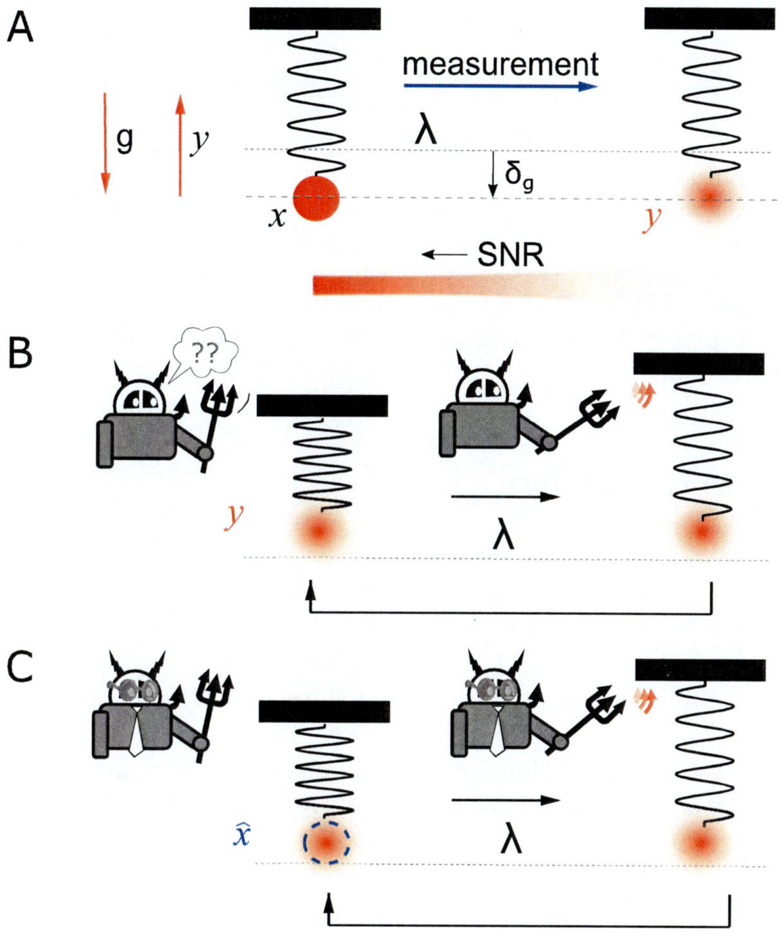

Fig. 6.1 Schematic information engine. (**a**) Noisy detector measures position y of the bead, actually located at x. Ratchet based on either (**b**) noisy position measurement y or (**c**) Bayesian position estimate $\hat{x}$ (blue dashed circle)

where $t_s = t'_s/\tau_r$ is the scaled time, $x_n \equiv x(nt_s)$ is the position of the bead at time step n, and ξ_n is the thermal noise, with $\langle \lambda_n \xi_n \rangle = 0$ and $\langle \xi_m \xi_n \rangle = \delta_{mn}$. The effect of measurement noise is modelled by a measurement variable y_n that is the sum of the bead's true position and additive white Gaussian noise,

$$y_n = x_n + \sigma_m \nu_n \,, \tag{6.2}$$

with ν_n a Gaussian random variable with zero mean and covariance $\langle \nu_k \nu_n \rangle = \delta_{kn}$. We also assume that the thermal noise affecting the bead's position is independent of the measurement noise: $\langle \xi_k \nu_n \rangle = 0$, for all k and n. The signal-to-noise ratio is defined as the amplitude ratio $\text{SNR} \equiv \sigma/\sigma_m$.

The position of the trap is updated using the feedback rule

$$\lambda_{n+1} = \lambda_n + \alpha(z_n - \lambda_n)\Theta(z_n - \lambda_n) , \tag{6.3}$$

where α is the feedback gain, Θ is the Heaviside function, and $z_n = \{y_n, \hat{x}_{n+1}\}$ is the estimate of the bead's position, using either the naive measurement y_n or the Bayesian estimate $\hat{x}_{n+1}$. The feedback gain α controls the input work done by the trap. For a high-SNR measurement followed by an instantaneous trap shift, the feedback gain $\alpha = 2$ would set the input work to zero. However, because of the feedback delay, the feedback gain α is tuned such that trap work is zero, as discussed in Chap. 4 (Fig. 4.3). In Sect. 6.3, we discuss the dependence of α on the measurement noise.

6.2 Energy Estimators

The stored gravitational energy during a ratchet, in the time interval $[t_n, t_{n+1})$, is given by $F_{n+1} = \delta_g\,(y_{n+1} - y_n)$, where δ_g is the scaled mass of the bead. The trap work increment arising from trap centre update is estimated as

$$W_{n+1} = \frac{1}{2}\left[(y_{n+1} - \lambda_{n+1})^2 - (y_{n+1} - \lambda_n)^2\right] , \tag{6.4}$$

where the energy estimated is based on the noisy measurements y. In Sect. 6.2.1, we show that Eq. (6.4) is an unbiased estimator of the true trap work.

6.2.1 *Unbiased Work Estimator*

In this section, we prove that the trap work calculated from the measured position y is an unbiased estimator of the true energy change, from stochastic thermodynamics, of the bead in the trap potential [13],

$$\mathcal{W}_{n+1} = \frac{1}{2}\left[(x_{n+1} - \lambda_{n+1})^2 - (x_{n+1} - \lambda_n)^2\right] . \tag{6.5}$$

Now we compare the work estimates to the true values of the thermodynamic quantities. Expanding the definition of the empirical input work using $y_k = x_k + \nu_k$, where ν_k is a zero-mean Gaussian random variable with variance σ_{m}^2, we obtain,

$$\begin{aligned} W_{n+1} &= \frac{1}{2}\left[(y_{n+1} - \lambda_{n+1})^2 - (y_{n+1} - \lambda_n)^2\right] \\ &= \frac{1}{2}\left[(x_{n+1} - \lambda_{n+1} + \nu_{n+1})^2 - (x_{n+1} - \lambda_n + \nu_{n+1})^2\right] \end{aligned}$$

$$= \frac{1}{2}\left[(x_{n+1} - \lambda_{n+1})^2 - (x_{n+1} - \lambda_n)^2\right] + (\lambda_{n+1} - \lambda_n)\,\nu_{n+1}$$

$$= \mathcal{W}_{n+1} + (\lambda_{n+1} - \lambda_n)\,\nu_{n+1}\,. \tag{6.6}$$

Averaging over the distribution of the noisy measurement at time step t_{n+1}, when the trap position is displaced $\lambda_{n+1} - \lambda_n \geq 0$, we find that

$$\begin{aligned}
\langle W_{n+1}\rangle_\nu &= \langle \mathcal{W}_{n+1}\rangle_\nu + \langle(\lambda_{n+1} - \lambda_n)\,\nu_{n+1}\rangle \\
&= \langle \mathcal{W}_{n+1}\rangle_\nu + \langle((\lambda_n + \alpha(z_n - \lambda_n) - \lambda_n)\,\nu_{n+1}\rangle \\
&= \langle \mathcal{W}_{n+1}\rangle_\nu + \alpha\,\langle z_n\,\nu_{n+1}\rangle - \alpha\,\langle \lambda_n\,\nu_{n+1}\rangle \\
&= \langle \mathcal{W}_{n+1}\rangle_\nu\,,
\end{aligned} \tag{6.7}$$

where, in the last step, we have used the property that both the position estimate z_n (which denotes the noisy measurement y_n in the naive case and the predictive estimate $\hat{x}_{n+1}$ in the Bayesian case), and the trap position λ_n at time step n are independent of the *future* measurement noise ν_{n+1}. Without the delay, the variables are correlated and the measurements are biased. This calculation suggests that feedback delay actually eliminates the bias in the empirical estimate of the trap work.

We also note that the trap work could alternatively be estimated using the Bayesian filtered position $\hat{x}_n$ in place of y_n in Eq. (6.4). However, the estimator in Eq. (6.4) is *robustly unbiased*: It is unbiased even if the dynamical model used in the Bayesian filter uses parameters that differ from those describing the physical system. The calculation of the trap position λ_n is based on past information and—whatever the algorithm used to compute its value—will always be independent of the current measurement.

6.3 Naive Information Engine

In this section, we discuss the operation of a naive information engine. The naive information engine operates by applying feedback directly based on the noisy measurements. The feedback rule used to update the trap position is

$$\lambda_{n+1} = \lambda_n + \alpha(y_n - \lambda_n)\Theta(y_n - \lambda_n)\,. \tag{6.8}$$

To operate the engine as a "pure" information engine, the feedback gain α is tuned such that $\dot{W} = 0$. For a feedback delay of one time step and SNR = 11, the critical feedback gain $\alpha^* \approx 1.5$, as shown in Fig. 6.2a. In addition to feedback delay (as discussed in Chap. 4), the measurement noise further reduces α^*, which becomes extremely small at low SNR, as shown in Fig. 6.2b. For SNR $\lesssim$ SNR$_c = 0.7 \pm 0.1$, where SNR$_c$ denotes the critical value of the SNR, a non-zero α^* could not be found

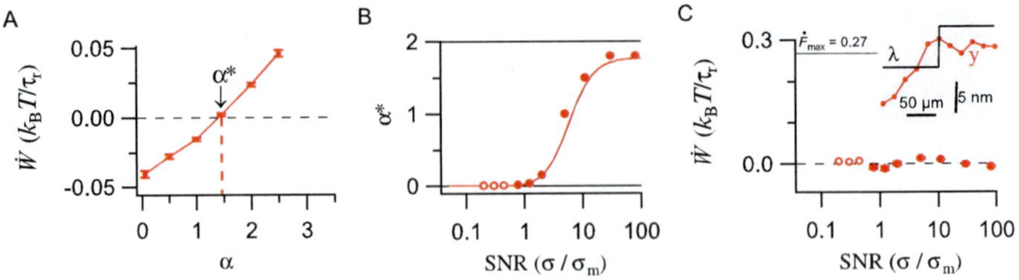

Fig. 6.2 Feedback gain as a function of SNR. Markers denote experimental data. (**a**) Trap power $\dot{W}$ as a function of the feedback gain α. The critical feedback gain $\alpha^* \approx 1.5$. (**b**) Critical feedback gain α^* for different SNR. Hollow markers denote the SNR values for which α^* could not be determined using the method outlined in (**a**). The solid line denotes the numerical-simulation results performed by Joseph Lucero. (**c**) Measured trap power for the critical feedback gain α^* presented in (**b**), as a function of SNR. The error bars (small compared to the markers) denote the standard error of the mean. (Inset) Experimental trajectory of the measured bead position (y, red) and trap position (λ, red) for $\alpha^* \approx 1.5$ and SNR = 11

experimentally using the procedure outlined in Fig. 6.2a. The experimental data for SNR $\lesssim$ SNR$_c$ are denoted by hollow red markers in Fig. 6.2b. In Sect. 6.4, we show that the vanishing of α^* corresponds to a kind of phase transition between a regime where one can set $\dot{F} > 0$ while maintaining $\dot{W} = 0$, and a regime where one cannot.

6.4 Phase Transition in Naive Information Engine

In this section, we present detailed experimental evidence for a phase transition in the naive information engine. Figure 6.3a shows the trap power as a function of the feedback gain α for different SNR, where the markers in the blue shaded region represent SNR values for which only positive trap powers ($\dot{W} > 0$) were measured. The experimental points are fit to a quadratic function; Fig. 6.3b shows the linear coefficients of the fits in Fig. 6.3a as a function of SNR. The measured slope switches from negative at high SNR $\gtrsim 0.8$ to positive at SNR $\lesssim 0.6$.

To measure the critical signal-to-noise ratio (where the linear coefficient is zero), the linear coefficients for SNRs 0.4–0.8 are linearly fit, giving an estimated zero linear coefficient and hence SNR$_c$ at 0.7 ± 0.1. In the naive-information-engine experiments, for SNR $<$ SNR$_c$, we cannot find an α that leads to $\dot{W} < 0$. Thus, the α^* we report in such cases (red hollow markers in Fig. 6.2) is an upper bound on the α^* that satisfies $\dot{W} \gtrsim 0$: Either there is a smaller positive α^* that we could not detect experimentally because differences in $\dot{W}$ are too close to zero, or the only value that (always) enforces $\dot{W} = 0$ is $\alpha^* = 0$. The corresponding $\dot{F}$ measured using the upper-bound value of α^* then is an upper bound on the output power for SNR<SNR$_c$.

The phase transition appearing in the naive information engine is removed by applying the feedback based on the Bayesian estimate of the bead's position. To

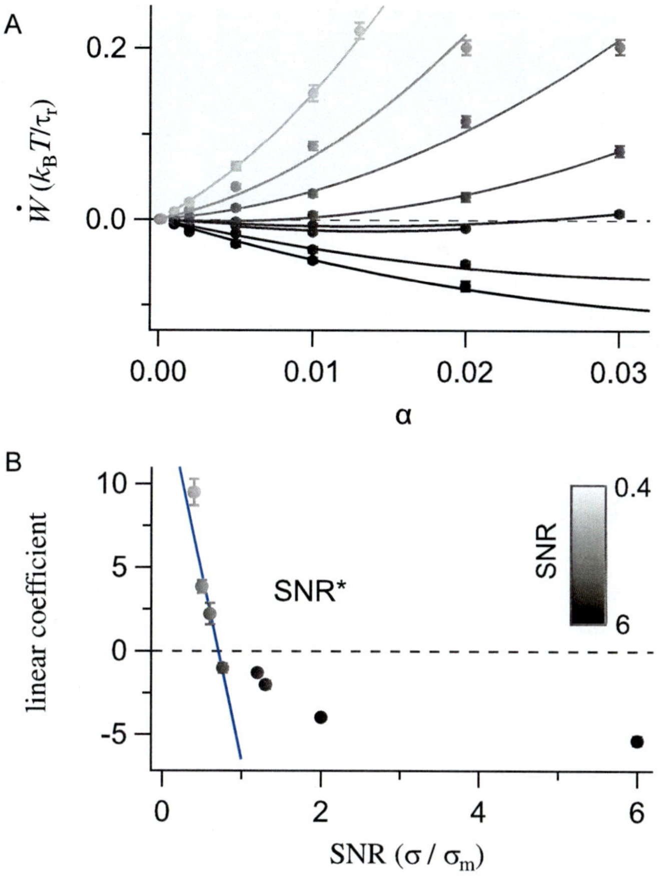

Fig. 6.3 Phase transition in naive information engine. (**a**) Trap power as a function of feedback gain α for signal-to-noise ratios 0.4, 0.5, 0.6, 0.8, 1.2, 1.3, 2, and 6 (different shades of grey, scale bar in (**b**)). Markers: experiment; curves: fit to quadratic functions. Blue shaded region: SNR values for which $\dot{W} > 0$ for all values of α. (**b**) Linear coefficients from quadratic fits (curves in (**a**)), as a function of SNR. Blue line: linear fit to first four points. Error bars each denote standard error of the mean in (**a**) and least-square parameter fit error of the linear coefficient in (**b**)

apply feedback based on the Bayesian filtered position estimate, we use the recursive algorithm discussed in Chap. 3. In Sect. 3.6, we presented experimental data to benchmark the performance of the Bayesian filter estimate.

6.5 Bayesian Information Engine

The Bayesian information engine applies feedback based on the filtered predictive estimate of the bead's position, given by

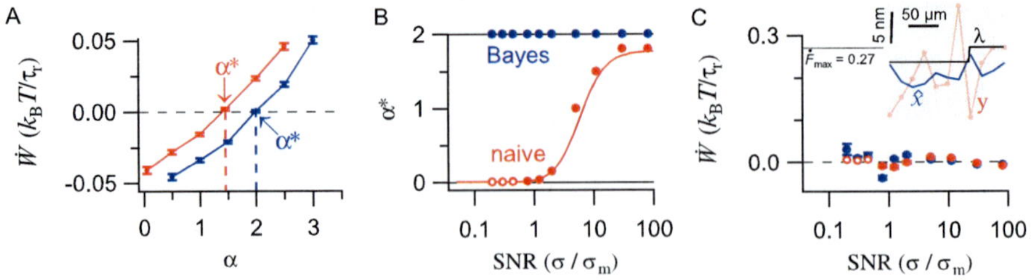

Fig. 6.4 Feedback gain as a function of SNR. Markers denote experimental data. (**a**) Trap power $\dot{W}$ as a function of the feedback gain α. The critical feedback gain $\alpha^* \approx 1.5$ for the naive (red) and $\alpha^* = 2$ for the Bayesian (blue) information engine. (**b**) Critical feedback gain α^* as a function of SNR for naive (red) and Bayesian (blue) information engines. The solid line denotes the numerical simulation result performed by Joseph Lucero (see [12], SI). (**c**) Measured trap power for the critical feedback gain α^* presented in (**b**), as a function of SNR. The error bars denote the standard error of the mean

$$\lambda_{n+1} = \lambda_n + \alpha(\hat{x}_{n+1} - \lambda_n)\Theta(\hat{x}_{n+1} - \lambda_n) \, . \tag{6.9}$$

As such, the Bayesian filter accounts in its internal model for feedback delays and measurement noise. Figure 6.4a shows the input trap power, at fixed SNR (=10), as a function of feedback gain α. Despite the delay and finite SNR, the input trap power is zero for the "ideal" feedback gain $\alpha^* = 2$. Figure 6.4c inset illustrates the operation of the Bayesian filter with a trajectory of the Bayesian information engine at lower SNR (= 2). In contrast to the naive information engine, the trap ratchets only when the estimated position (blue) crosses the trap centre (black), and not necessarily when the noisy measurement (light red) crosses the trap centre.

In contrast to the naive engine, the critical feedback gain α^* remains near 2 for the Bayesian engine for all SNR values, as shown in Fig. 6.4b. The corresponding measured input trap powers for both the naive and Bayesian information engines are close to zero relative to the maximum output power ($\dot{F}_{\max} \approx 0.27\, k_B T/\tau_r$) of the engine, at all SNR [Fig. 6.4c]. Figure 6.5 shows that $\frac{d\dot{W}}{d\alpha}\Big|_{\alpha=0}$ is negative for the Bayesian information engine for all SNR. This implies that there is always a positive feedback gain ($\alpha = 2$) that sets $\dot{W} = 0$. Thus, the Bayesian information engine eliminates the phase transition, and the critical feedback gain α^* can be always found using the method outlined in Fig. 6.4a. By contrast, for a naive information engine, the linear coefficient ($\equiv \frac{d\dot{W}}{d\alpha}$) becomes positive below SNR_c [Fig. 6.3b].

Taking advantage of predictions in our estimation algorithm thus simplifies the experiments, as it eliminates the need to empirically tune the feedback gain, ensuring that the zero-work condition is always satisfied at $\alpha = 2$. Above, we saw that it also simplifies the work calculations needed to realize a pure information engine, as the value calculated directly from the noisy measurement is an unbiased estimator of the true work.

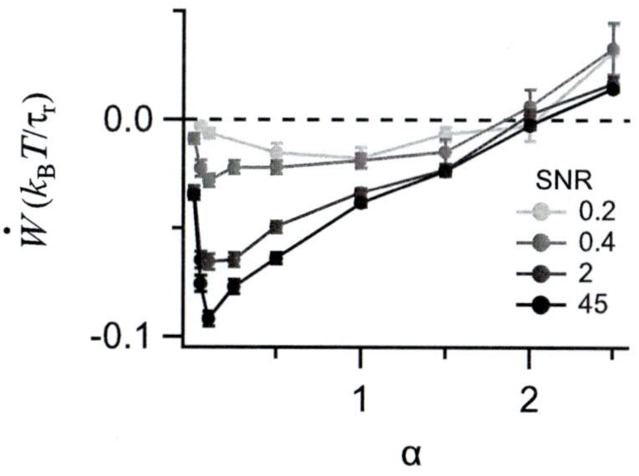

Fig. 6.5 Trap power $\dot{W}$ as a function of feedback gain α for different signal-to-noise ratios (different shades of grey). Solid curves connect solid markers to ease visualization. Error bars denote the standard error of the mean

6.6 Results

Finally, Fig. 6.6a compares the performance of the naive and Bayesian information engines, as quantified by the rate of stored gravitational power $\dot{F}$, while keeping the rate of trap work $\dot{W} = 0$. Both output powers $\dot{F}$ increase monotonically with SNR and saturate at the same power level at high SNR (> 10). The output power as a function of SNR for the naive information engine (red) has different slopes on either side of SNR_c. This non-analytic behaviour of $\dot{W}(\text{SNR})$ at SNR_c is the signature of a phase transition in the model describing ratchet output.

Although the performance of Bayesian and naive information engines is similar at low and high SNR, there is a striking contrast at intermediate SNR ($\lesssim 1$). Indeed, the difference of output powers ($\dot{F}_\text{B} - \dot{F}_\text{N}$), normalized by $\dot{F}_\text{max}$, significantly exceeds zero for $0.1 \leq \text{SNR} \leq 2$ and reaches a maximum at $\text{SNR} \approx \text{SNR}_c$ (Fig. 6.6b).

At high SNR, the Bayesian filter "trusts the observation" and returns an estimate close to the instantaneous measurement, corrected for the expected bias due to the time delay. Since this bias is small for frequent measurements, both engines have similar performance and take advantage of all favourable thermal fluctuations, saturating at the maximum output power $\dot{F}_\text{max} \approx 0.27$. At low SNR, the measurements are so noisy that they exceed the scale of the trap. The Bayesian information engine then extracts negligible power, while the naive engine extracts zero power. Therefore, at $\text{SNR} \gg 1$ and $\ll 1$, the difference of output powers ($\dot{F}_\text{B} - \dot{F}_\text{N}$) tends to zero. But at intermediate SNR, the effective noise averaging in the Bayesian (Kalman) filter produces more accurate estimates, leading to better feedback decisions and thus improved engine performance.

We find that the phase transition arises from the *biased* estimate of the bead's position from the noisy measurements. This bias has two origins: the delay due to feedback latency and the failure of the naive measurement to account for the fact that fluctuations above threshold are rare while fluctuations of either sign are equally likely. Because fluctuations up to the threshold are rare, the bead is usually below the observed value whenever an apparent threshold crossing is observed.

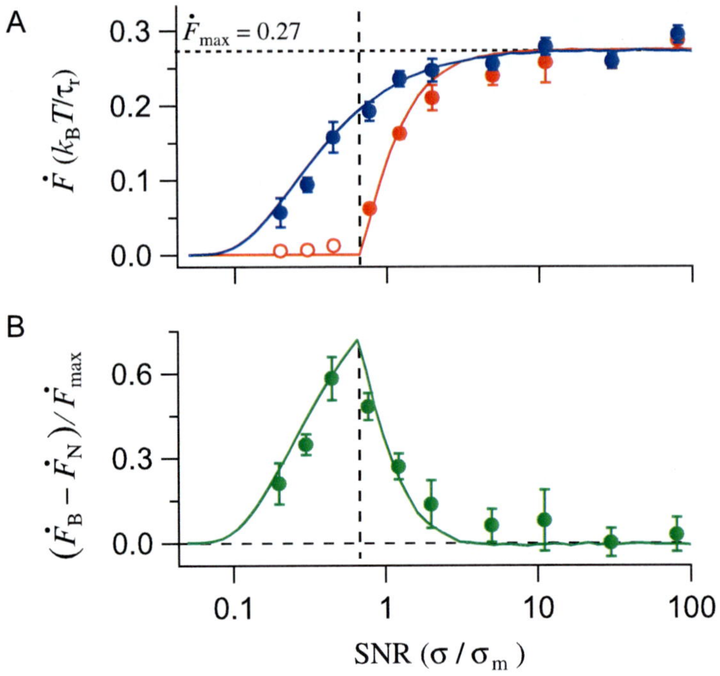

Fig. 6.6 Performance of the information engines. (**a**) Output power of naive (red) and Bayesian (blue) information engines as a function of SNR. Hollow red markers denote output power at $\alpha^* = 0$. (**b**) Difference of output work extraction rates for the Bayesian ("B") and naive ("N") engines scaled by the maximum rate ($\dot{F}_{max} = 0.27$). The difference peaks at SNR=SNR$_c$ ≈ 0.7 (vertical dashed lines). Markers denote experimental means, solid curves the numerical simulations performed by Joseph Lucero. Error bars denote (**a**) standard error of the mean and (**b**) propagated standard error of the mean from (**a**)

By contrast, a phase transition does *not* occur for the Bayesian information engine. The Bayesian filter gives an *unbiased* estimate of the bead's position using the "prior" knowledge of observing the bead near the threshold, based on the model of the system dynamics (Eq. 3.10). The Bayesian filter also accounts for the feedback delay and predicts the position of the bead (Sect. 3.6). As a result, the bead is equally likely to be on either side of the predicted position, allowing one to tune for zero trap power and extract at least some power at *any* SNR value (Sect. 6.5).

6.7 Conclusion

Information engines that decide whether to ratchet using single noisy measurements have a phase transition at a critical signal-to-noise ratio SNR$_c$ and cannot function for SNR $<$ SNR$_c$. By contrast, if its feedback uses an unbiased Bayesian estimate of bead position that incorporates prior measurements, an information engine can operate at *all* values of SNR. The maximum performance benefit over the naive engine occurs at the critical value SNR_c. In Appendix A, we show that feedback based on Bayesian estimate of the bead position also improves the performance of information engines that are designed to generate translation (the transporter engine) and to cool the thermal bath.

The ability to increase the performance of an information engine at low SNR is important for experimental investigations of motor mechanisms that use fluorescent probes [14]. In such applications, lower light intensities for monitoring fluorescent probes reduce photobleaching and allow longer measurements of motor behaviour.

In addition, using a filtering algorithm to reduce the required accuracy of information while maintaining a given performance may decrease the thermodynamic costs of processing position measurements. Generally, a lower measurement accuracy reduces the minimum thermodynamic costs of running the controller [3, 15]. However, keeping a memory of past observations should increase those costs. Further work is needed to understand the efficiency of a feedback strategy that incorporates a memory of past measurements with one based purely on the most recent measurement. Such studies could evaluate the potential performance tradeoffs encountered when varying the measurement accuracy.

References

1. T.K. Saha, J.N.E. Lucero, J. Ehrich, D.A. Sivak, J. Bechhoefer, Maximizing power and velocity of an information engine. Proc. Natl. Acad. Sci. U.S.A **118**, e2023356118 (2021)
2. T.K. Saha, J. Bechhoefer, Trajectory control using an information engine. Opt. Trapp. Opt. Micromanipul. XVIII **11798**, 117980L (2021)
3. G. Paneru, S. Dutta, T. Sagawa, T. Tlusty, H.K. Pak, Efficiency fluctuations and noise induced refrigerator-to-heater transition in information engines. Nat. Commun. **11**, 1–8 (2020)
4. A. Taghvaei, O.M. Miangolarra, R. Fu, Y. Chen, T.T. Georgiou, On the relation between information and power in stochastic thermodynamic engines. IEEE Control Syst. Lett. **6**, 434–439 (2022)
5. K. Nakamura, T.J. Kobayashi, Connection between the bacterial chemotactic network and optimal filtering. Phys. Rev. Lett. **126**, 128102 (2021)
6. J.M. Horowitz, T. Sagawa, J.M.R. Parrondo, Imitating chemical motors with optimal information motors. Phys. Rev. Lett. **111**, 010602 (2013)
7. N. Rupprecht, D.C. Vural, Predictive Maxwell's demons. Phys. Rev. E **102**, 062145 (2020)
8. A.P. Fields, A.E. Cohen, Electrokinetic trapping at the one nanometer limit. Proc. Natl. Acad. Sci. U.S.A. **108**, 8937–8942 (2011)
9. L. Magrini, et al., Real-time optimal quantum control of mechanical motion at room temperature. Nature **595**, 373–377 (2021)
10. G.P. Conangla, et al., Optimal feedback cooling of a charged levitated nanoparticle with adaptive control. Phys. Rev. Lett. **122**, 223602 (2019)
11. J.N.E. Lucero, J. Ehrich, J. Bechhoefer, D.A. Sivak, Maximal fluctuation exploitation in Gaussian information engines. Phys. Rev. E **104**, 044122 (2021)
12. T.K. Saha, J.N.E. Lucero, J. Ehrich, D.A. Sivak, J. Bechhoefer, Bayesian information engine that optimally exploits noisy measurements. Phys. Rev. Lett. **129**, 130601 (2022)
13. K. Sekimoto, *Stochastic Energetics* (Springer, Berlin, 2010)

14. C. Veigel, C.F. Schmidt, Moving into the cell: single-molecule studies of molecular motors in complex environments. Nat. Rev. Mol. Cell Biol. **12**, 163–176 (2011)
15. J.M. Horowitz, H. Sandberg, Second-law-like inequalities with information and their interpretations. New J. Phys. **16**, 125007 (2014)

Chapter 7
Information Engine in a Nonequilibrium Bath

Maxwell's famous thought experiment, which converted information about thermal fluctuations into energy, has improved our understanding of the second law of thermodynamics [1]. These engines, now known as information engines, have recently been experimentally realized in many physical systems, with a focus on demonstrations of function and on their efficiencies [2–7]. To be consistent with the second law of thermodynamics, the information-processing cost must equal or exceed the energy extracted by the information engine [8–11].

In this Chapter, we present an information engine that is coupled to a medium that is at an equilibrium temperature T but also exhibits external noise beyond what is expected from the thermal fluctuations. Such a bath acts as an energy source with contributions from both thermal and nonequilibrium (external) fluctuations. We show that an information engine can extract significantly more energy from this type of nonequilibrium bath than it can from an equilibrium thermal bath—up to ten times more power than from a bath at room temperature, for the nonequilibrium baths that we are able to create.

Contribution to the Chapter: The results presented in this chapter have been published in *Physical Review Letters* [12]. John Bechhoefer suggested the project. I performed the experiments and analyzed the data presented in this chapter. Momčilo Gavrilov and I modified the sample chamber to add the external electric forces. I characterized the noise source and frequency response of the amplifier. Jannik Ehrich performed the numerical simulations and analytic calculations. All authors contributed to writing the paper.

7.1 Experimental Apparatus

Our information engine consists of an optically trapped heavy bead in a water medium that acts as a thermal bath at room temperature T. The Johnson noise

T. K. Saha, *Information-Powered Engines*, Springer Theses,
https://doi.org/10.1007/978-3-031-49121-4_7

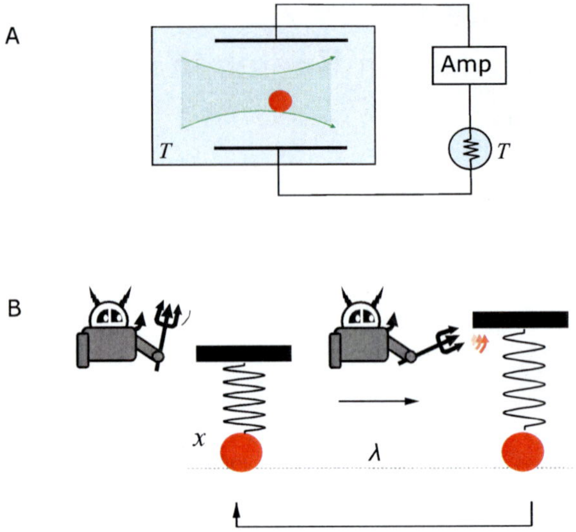

Fig. 7.1 Schematic of information engine in a non-equilibrium bath. (**a**) Optically trapped bead, in a water medium, in contact with a nonequilibrium external noise source. The external noise is generated by the electrodes that are connected to a resistor. Both bead and resistor are at room temperature T. (**b**) Schematic of the information engine

of a 300 kΩ resistor, also at room temperature T, is amplified and filtered to create the nonequilibrium forcing [13]. The resistor is "connected" to the bead using electrodes, as shown in Fig. 7.1a. The active-noise strength is regulated using one variable- and two fixed-gain amplifiers. The total electronic noise from the amplifiers is much smaller than the Johnson noise of the resistor. Under gravity, the bead fluctuates about an *equilibrium* (mean) position because of thermal and active forces acting on it. We measure the bead position at a sampling time $t_s = 20\,\mu\text{s}$, using the forward-scattered light from a detection laser.

We model the trapped bead as a spring-mass system, as shown in Fig. 7.1b. The information engine operates by raising the trap position when the bead fluctuates above the trap centre. The upward fluctuation of the bead, increasing its gravitational potential energy, is rectified by ratcheting the trap position. We carefully update the trap position to convert the fluctuations from the nonequilibrium bath into gravitational energy without any contributions from the trap.

7.2 Equations of Motion

The equation of motion of an optically trapped bead in an active bath is described by the Langevin equation

$$\dot{x}(t) = \underbrace{-\,(x(t) - \lambda(t))}_{\text{restoring force}} - \underbrace{\delta_g}_{\text{grav. force}} + \underbrace{\sqrt{2}\,\xi(t)}_{\text{thermal noise}} + \underbrace{\sqrt{2D_{\text{ne}}}\,\zeta(t)}_{\text{nonequilibrium noise}}\,, \tag{7.1}$$

where $x(t)$ is the bead position at time t, with the scaled mass $\delta_g = \Delta m\, g/\kappa\sigma$, in a trap of stiffness κ centred at λ and σ the standard deviation of the bead position in the static trap. The length and time in Eq. (7.1) have been scaled, similar to Eq. (2.12). The thermal noise term $\xi(t)$ is modelled by a Gaussian random variable with zero mean and variance $\langle \xi(t)\xi(t') \rangle = \delta(t - t')$.

The nonequilibrium noise $\zeta(t)$ is generated by applying electric voltages to the bead, using the apparatus described in Sect. 3.8. The bead experiences electrokinetic forces, which are a combination of electroosmotic and electrophoretic forces [14], because of the electric field applied on the bead. The strength of the nonequilibrium force is characterized by D_{ne}, which is scaled by the ordinary diffusivity D of the bead. The electric forces are generated by amplifying and low-pass filtering the Johnson noise of a resistor and can be modelled by

$$f_{ne}^{-1}\, \dot{\zeta}(t) = -\zeta(t) + \xi'(t)\,, \tag{7.2}$$

where f_{ne} is the corner frequency of the noise, which is set by the low-pass filter, and where the thermal noise $\xi'(t)$ in the resistor follows $\langle \xi'(t) = 0 \rangle$ and $\langle \xi'(t)\xi'(t') \rangle = \delta(t - t')$.

The position of the trap is updated, at discrete time intervals of $t_s = 20\,\mu\text{s}$, using the feedback algorithm

$$\lambda_{n+1} = \lambda_n + \alpha\,(x_n - \lambda_n)\,\Theta(x_n - \lambda_n)\,, \tag{7.3}$$

where $\Theta(\cdot)$ is the Heaviside (step) function, and α is the feedback gain, which controls the input trap power $\dot{W}$. The trap power can be set to zero at $\alpha = 2$ in the absence of feedback delay [15]. However, in the presence of feedback delay $t_d = t_s = 20\,\mu\text{s}$, the zero-work condition is realized at lower feedback gain $\alpha \approx 1.8$, when the measurement noise is small [16]. The signal-to-noise ratio SNR= 20 is high enough that the effects of measurement noise can be neglected.

7.3 Energy Measurements

The stored gravitational energy (more formally, the *free energy*) of the bead is

$$\Delta F_{n+1} = \delta_g\ (\lambda_{n+1} - \lambda_n)\ . \tag{7.4}$$

The output power is measured using $\dot{F} = \left\langle \left(\sum_{k=0}^{N} \Delta F_k \right) / N t_s \right\rangle$, where $\langle \cdot \rangle$ denotes the average over multiple trajectories, each of length N time-steps. The work done by the trap on the bead is

$$W_{k+1} = \frac{1}{2}\left[(x_{n+1} - \lambda_{n+1})^2 - (x_{n+1} - \lambda_{n+1})^2 \right], \tag{7.5}$$

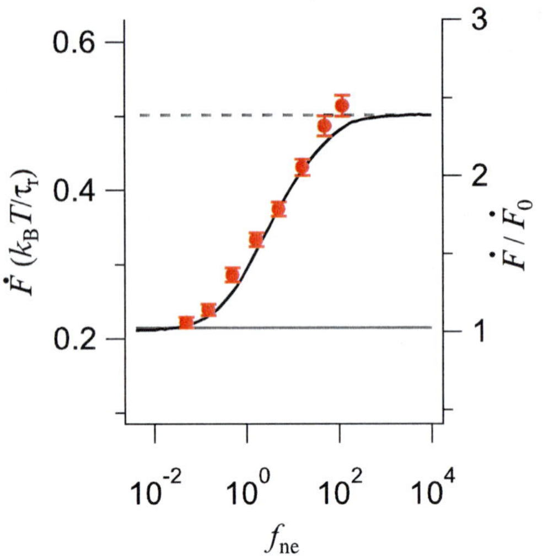

Fig. 7.2 Output power as a function of corner frequency of the nonequilibrium bath f_{ne} scaled by the trap's corner frequency f_r, at $D_{ne} = 3.0$. The grey line denotes the measured output power at $D_{ne} = 0$ (the usual thermal bath). The black curve denotes numerical simulations

where positive W_{k+1} indicates that work is done by the trap on the bead. Similarly, the trap power is measured using $\dot{W} = \left\langle \left(\sum_{k=0}^{N} \Delta F_k \right) / N t_s \right\rangle$.

7.4 Frequency Dependence of Output Power

We study the dependence of the output power of the information engine on the characteristics of the nonequilibrium noise. First, we measure the output power at different corner frequencies f_{ne} of the nonequilibrium noise source. We fix the amplitude of the nonequilibrium noise to $D_{ne} = 3.0$ and vary the nonequilibrium noise corner frequency f_{ne}, by changing the time constant of the low-pass filter that is applied to the resistor Johnson noise. Note that the corner frequency is expressed in units of the corner frequency of the trap, $f_c = 200 \pm 2$ Hz. We find the output power $\dot{F}$ increases with f_{ne}, as shown in Fig. 7.2. The output power saturates for $f_{ne} \gtrsim 100$. At high f_{ne}, the nonequilibrium noise acts as Gaussian white noise, and the bead experiences a thermal bath with a higher "effective temperature". At low $f_{ne} \lesssim 10^{-2}$, the fluctuations from the thermal noise dominate the low-frequency nonequilibrium noise. The right-axis represents the ratio of the output power $\dot{F}$ and F_0 (output power at $D_{ne} = 0$). At the maximum f_{ne}, with fixed $D_{ne} = 3.0$, the output power is four times the power extracted from the thermal bath (for $D_{ne} = 0$).

7.5 Amplitude Dependence of Output Power

Next, we study the dependence of the output power on the strength D_{ne} of the nonequilibrium noise. We varied D_{ne} using a variable-gain amplifier. We fixed the

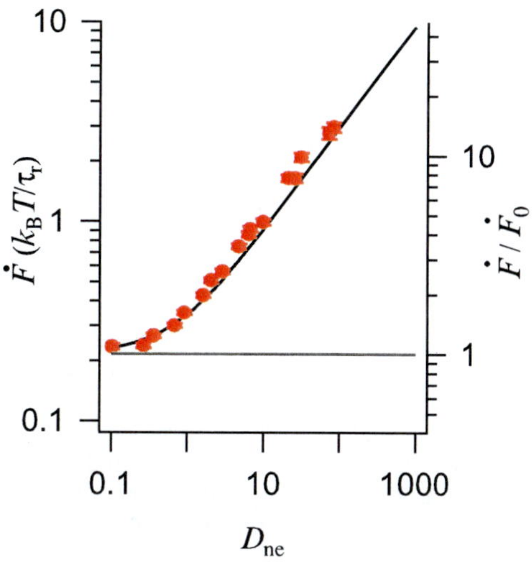

Fig. 7.3 Output power as a function of the strength of the nonequilibrium noise. The grey line denotes the measured output power at $D_{ne} = 0$ (usual thermal bath). The black curve denotes the theory plot, calculated by Jannik Ehrich. For $D_{ne} \gg 1$, the output power $\dot{F} \sim D_{ne}^{1/2}$

corner frequency at the maximum $f_{ne} \approx 100$ and feedback gain $\alpha = 1.8$. We find that the output power increases with the nonequilibrium-noise strength D_{ne}, as shown in Fig. 7.3 (left-axis). The power saturates at low D_{ne} to that extracted from a thermal bath at room temperature. Away from that limit, the output power increases monotonically with D_{ne}. In our experiments, the amplifier gain limits the maximum achievable noise strength and hence the output power of the information engine. At the maximum corner frequency f_{ne} and strength D_{ne} of the nonequilibrium noise, we achieve a ten-fold increase in the output power of the information engine compared to the equilibrium thermal bath (Fig. 7.3, right-axis).

7.6 Optimum Scaled Mass

Finally, we study the dependence of the output power on the scaled mass δ_g. For $D_{ne} = 0$, represented in black in Fig. 7.4, the output power is maximized at the optimum scaled mass $\delta_g^* = 0.8$, as seen in Chap. 4. The optimum arises from the balance between the gain in gravitational energy (increasing the output) and the decrease in rate of favourable thermal fluctuations. For a heavy scaled mass, the mean position of the bead $(-\delta_g)$ is farther away from the trap centre. Thus, every ratchet event stores a large amount of gravitational energy; however, such fluctuations become increasingly rare. At higher values of the nonequilibrium noise, the optimum δ_g shifts towards a heavier scaled mass, as shown in Fig. 7.4. The nonequilibrium noise increases the amplitude of the bead's fluctuation and makes barrier crossing for a heavier mass more frequent. As heavier masses store more energy, a higher frequency of ratchet events in the presence of nonequilibrium shifts the optimum to a higher δ_g.

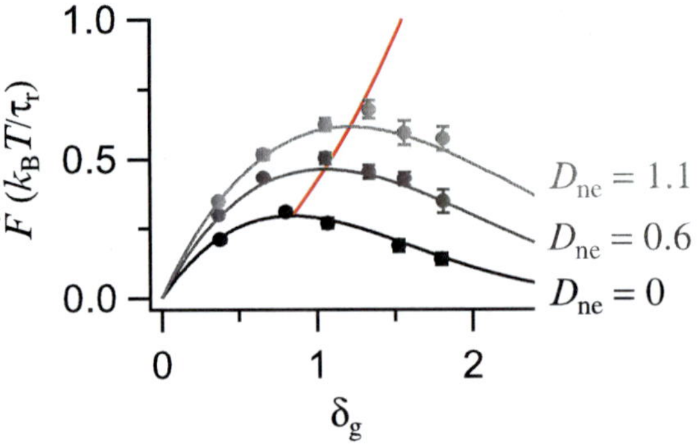

Fig. 7.4 Optimizing scaled mass. The markers denote the output power as a function of the scaled mass for different strength of the nonequilibrium noise: $D_{ne} = 0$ (black), $D_{ne} = 0.6$ (grey), and $D_{ne} = 1.1$ (light grey). The red solid line denotes the optimum scaled mass δ_g^* at different D_{ne}. The markers denotes the experimental data, and the lines denote theory plots, calculated by Jannik Ehrich

For every nonequilibrium noise strength D_{ne} there is an optimum scaled mass δ_g^* that maximizes the output $\dot{F}$ of the engine. The red line in Fig. 7.4 represents the maximum achievable output power $\dot{F}$ for different nonequilibrium noise strengths D_{ne}, evaluated at the optimum scaled mass δ_g^*.

7.7 Conclusion

Information engines that operate by exploiting thermal fluctuations in an equilibrium bath are limited by the internal temperature T of the medium. In Chap. 4, we found that the maximum power that the information engine can extract is 0.295 k_BT/τ_r. Lucero et al. showed that there are no feedback algorithms that can extract more power, which are both affine transformations of our feedback algorithm and satisfy zero input work [17]. More loosely, no feedback algorithm can extract more power than $\sim 1\ k_BT/\tau_r$, bounded by the natural rate of fluctuations that a system experience in a thermal bath at temperature T.

We showed that our information engine can extract energies from a nonequilibrium bath that are far larger than the thermal-bath bound. Indeed, the theoretical analysis of Jannik Ehrich and our results in Fig. 7.3 suggest that the maximum power is increased by a factor $\sqrt{D_{ne}/D}$ relative to the ordinary power k_BT/τ_r that can be extracted from a thermal bath. The additional noise in the surrounding acts as a higher "effective" temperature for the information engine. Thus, the engine operates in a bath which acts as if it were at a higher temperature than the measuring device, the FPGA (which is at room temperature). Previous studies have predicted that information engines that operate in a bath at a higher temperature than the measuring device can extract net positive work [18]. Although the focus of the study presented in this chapter was on maximizing the output performance, measuring the information-processing cost is also of interest in the case of a nonequilibrium bath. The engine can use the energy present in non-thermal fluctuations of the bath to extract more energy than it dissipates to process information, which would make the information-to-energy efficiency > 1. Note that this is not the thermodynamic efficiency, which is bounded by unity. However, further investigation is needed to measure the information-processing cost to quantify efficiency.

In our experiments, we build the nonequilibrium bath by amplifying the Johnson noise of a resistor. We do not account for the cost of generating the external noise, because we envision operating the information engine in a naturally occurring nonequilibrium environment, where one does not have to spend energy to maintain the nonequilibrium bath. For example, molecular motors operate in a medium that is out of equilibrium, with extra fluctuating forces provided by the motion of surrounding motors. Such motion leads to hydrodynamic flow in a fluid medium, which produces forces that can be rectified by a molecular motor operating as an information ratchet. We will discuss this in more detail in Chap. 8. Other examples include sailboats, which use the motion of the wind to move in any direction by rectifying the wind flow. In such a situation, it is clear that one does not have to maintain the nonequilibrium bath, and energy can be extracted for "free".

Although information engines and sailboats operate at vastly different length scales, they have the same working principle—rectify favourable fluctuations. The performance of both, information engines and sailboats, are limited by the characteristics of the external noise and the applied feedback. With a global shortage in power supply, energy harvesting from renewable sources has become important. Research on information engines has led to a better understanding of the feedback designs that maximize both power output [15–17] and energy efficiency [5, 6, 19]. The optimization principles found for information engines might help improve the performance of energy-harvesting engines such as sailboats [20], wind turbines [21, 22], self-winding watches [23], and circuit rectifiers [24].

References

1. H. Leff, A.F. Rex, *Maxwell's Demon 2: Entropy, Classical and Quantum Information, Computing* (CRC Press, Boca Raton, 2002)
2. S. Toyabe, T. Sagawa, M. Ueda, E. Muneyuki, M. Sano, Experimental demonstration of information-to-energy conversion and validation of the generalized Jarzynski equality. Nat. Phys. **6**, 988–992 (2010)
3. J.V. Koski, V.F. Maisi, J.P. Pekola, D.V. Averin, Experimental realization of a Szilard engine with a single electron. Proc. Natl. Acad. Sci. U.S.A. **111**, 13786–13789 (2014)
4. K. Chida, S. Desai, K. Nishiguchi, A. Fujiwara, Power generator driven by Maxwell's demon. Nat. Commun. **8**, 1–7 (2017)
5. G. Paneru, D.Y. Lee, T. Tlusty, H.K. Pak, Lossless Brownian information engine. Phys. Rev. Lett. **120**, 020601 (2018)
6. M. Ribezzi-Crivellari, F. Ritort, Large work extraction and the Landauer limit in a continuous Maxwell demon. Nat. Phys. **15**, 660–664 (2019)

7. N. Cottet, et al., Observing a quantum Maxwell demon at work. Proc. Natl. Acad. Sci. U.S.A **114**, 7561–7564 (2017)
8. C.H. Bennett, The thermodynamics of computation-a review. Int. J. Theor. Phys. **21**, 905–940 (1982)
9. J.M.R. Parrondo, J.M. Horowitz, T. Sagawa, Thermodynamics of information. Nat. Phys. **11**, 131–139 (2015)
10. D. Mandal, C. Jarzynski, Work and information processing in a solvable model of Maxwell's demon. Proc. Natl. Acad. Sci. U.S.A. **109**, 11641–11645 (2012)
11. J.M. Horowitz, H. Sandberg, Second-law-like inequalities with information and their interpretations. New J. Phys. **16**, 125007 (2014)
12. T.K. Saha, et al., Information Engine in a Nonequilibrium Bath. Phys. Rev. Lett. **131**, 057101 (2023)
13. J.B. Johnson, Thermal agitation of electricity in conductors. Phys. Rev. **32**, 97 (1928)
14. A.E. Cohen, Trapping and manipulating single molecules in solution Ph.D. Thesis, Stanford University (2007)
15. T.K. Saha, J.N.E. Lucero, J. Ehrich, D.A. Sivak, J. Bechhoefer, Maximizing power and velocity of an information engine. Proc. Natl. Acad. Sci. U.S.A **118**, e2023356118 (2021)
16. T.K. Saha, J.N.E. Lucero, J. Ehrich, D.A. Sivak, J. Bechhoefer, Bayesian information engine that optimally exploits noisy measurements. Phys. Rev. Lett. **129**, 130601 (2022)
17. J.N.E. Lucero, J. Ehrich, J. Bechhoefer, D.A. Sivak, Maximal fluctuation exploitation in Gaussian information engines. Phys. Rev. E **104**, 044122 (2021)
18. S. Still, Thermodynamic cost and benefit of memory. Phys. Rev. Lett. **124**, 050601 (2020)
19. T. Admon, S. Rahav, Y. Roichman, Experimental realization of an information machine with tunable temporal correlations. Phys. Rev. Lett. **121**, 180601 (2018)
20. Y. An, J. Yu, J. Zhang, Autonomous sailboat design: a review from the performance perspective. Ocean Eng. **238**, 109753 (2021)
21. F. Porté-Agel, M. Bastankhah, S. Shamsoddin, Wind-turbine and wind-farm flows: a review. Bound.-Layer Meteorol. **174**, 1–59 (2020)
22. R.S. Amano, Review of wind turbine research in 21st century. J. Energy Resour. Technol. **139**(5), 050801 (2017). https://doi.org/10.1115/1.4037757
23. V. Leonov, *Wearable Monitoring Systems* (Springer, Berlin, 2011), pp. 27–49
24. T. Lehmann, Y. Moghe, On-chip active power rectifiers for biomedical applications, in *2005 IEEE International Symposium on Circuits and Systems* (2005), pp. 732–735

Chapter 8
Identifying Information Engines

Information engines operate to convert fluctuations in the surrounding medium into gravitational potential energy [1], generate directed motion [2] or just cool their surroundings [3]. These engines can extract energy even from nonequilibrium baths (Chap. 7). Such nonequilibrium settings occur naturally inside cells, where molecular motors operate. In Chap. 7, we found that information engines can produce higher output in a nonequilibrium bath. Although molecular motors such as kinesin are generally believed to operate by fluctuating-ratchet mechanisms that are independent of system state, it is possible that, at least in some cases, they operate as an information ratchet [4]. And perhaps some other molecular motors in the cell do, too. In this chapter, I investigate how one might recognize an information engine (ratchet) by observing its behaviour as it operates and contrast it against ordinary ratchets. The difference between the two cases is that ordinary ratchets fluctuate and generate directed motion, but the ratchet events are independent of the system state. For an information ratchet, the ratchet events are correlated with system fluctuations. Is it possible to distinguish the two situations empirically? Information engines have been viewed as a "textbook example" to understand the intricacies of the second law. However, as motivated in this thesis, information engines can also produce large amounts of energy and generate significant translation speeds, and can be of practical use in the real world. If information engines naturally exist, then understanding their design principles could open the research field to build self-contained microscopic information engines with application in biotechnology.

In this Chapter, I seek methods to distinguish the physical aspects of the information engine from a power-stroke (heat) engine. In general, the operation of the molecular motor could be very complicated. A simple power-stroke mechanism, used to model molecular motors, involves relaxation of a constrained spring that is triggered by a chemical reaction [5]. Here the term "power stroke" is used to refer to a slightly different mechanism, one where the spring applies a constant (average) force: The motor simply moves the trap centre at constant speed, so that the bead on average moves at the same speed. To isolate the differences between the two kinds

T. K. Saha, *Information-Powered Engines*, Springer Theses,
https://doi.org/10.1007/978-3-031-49121-4_8

of engines, I use the same experimental apparatus. I measure the energies flowing in the power-stroke engine and find that a power-stroke engine dissipates heat into the medium because of friction between the bead and the viscous medium—water [6]. I also consider a "drift-diffusion" motor, which adds random forcing to the steady power-stroke motion. Again, heat is dissipated into the medium. By contrast, a translating bead powered by an information engine does not dissipate any heat; rather, it extracts heat from the medium and converts it into useful energy. That heat is absorbed by an information engine while it translates through a medium distinguishes it from other power-stroke engines.

The rate of heat dissipation produced by a moving bead is of the order 10^{-18} W. Measuring such small energy rates, localized near the bead, is experimentally challenging. The stochastic thermodynamics techniques discussed in Chap. 2 can be used to infer the heat dissipated by the bead [7]. However, it requires the knowledge of controller model and state—in this case, the position of the harmonic trap. I show that the information engine can be easily distinguished from other engines if such information is available.

Heat dissipated by a power-stroke engine arises from Stokes drag, which also leads to hydrodynamic flow in the medium. I measure the heat dissipated by the power-stroke engines and show that it follows the Stokes predictions. Stokes dissipation also leads to hydrodynamic flow. However, the absence of Stokes heat dissipation in information engines has a negligible effect on the hydrodynamic flow, as discussed in Appendix B.

Finally, I measure the probability distribution of the bead trajectories for the two engines. I find that the engines can be distinguished using only the bead position distributions, without knowledge of experimental parameters such as the trap stiffness κ, trap position λ, etc. A similar principle can be implemented to identify molecular motors operating as information engines, by measuring for example the position of the cargo attached to a kinesin motor, without any knowledge of the function or state of the molecular motor.

Contribution to the Chapter: The studies presented in this chapter have not yet been submitted for publication. I performed the experiments presented in this chapter. John Bechhoefer and I proposed the research idea, designed the experiments, analyzed the data, and wrote the paper.

8.1 Introduction

Figure 8.1 contrasts the (a) power-stroke and (b) information engines. Let's consider two ways to raise a microscopic bead, attached by a spring, against gravity in a water medium (acting as a thermal bath). Because of thermal fluctuations, the bead fluctuates about a mean position. In Fig. 8.1a, the climber (operating as a power-stroke engine) does work to raise the bead. Because of viscous dissipation, a significant fraction of the work done by the climber is lost as heat to the surrounding fluid.

Fig. 8.1 Schematic of power-stroke (left) and information (right) engines that raise beads in a potential landscape

By contrast, the climber operating as an information engine raises the bead position by extracting the fluctuations in the thermal bath (Fig. 8.1b). The intelligent climber "sits" on the bead and fluctuates along with it, tethering itself to the current energy step by a spring-attached trident. When the bead fluctuates up to the next step, the climber grabs onto the next step, therefore biasing up the motion of the bead. The engine thus converts heat from the bath into gravitational potential energy. The climber does not do work on the bead, and instead makes use of the information about the upward fluctuation of the bead. Because the accompanying information processing necessarily dissipates at least as much as the work extracted, the second law of thermodynamics holds for the complete system of engine plus measuring device plus controller.

8.2 Power-Stroke and Information Controller

I experimentally realize the power-stroke and information engines using an optically trapped bead in water. I use identical experimental apparatus (described in Chap. 3) to operate both engines; the only difference is the algorithm that updates the position of the trap. For a power-stroke engine, the trap position is updated using

$$\lambda_{n+1} = \lambda_n + c\,, \tag{8.1}$$

where c is a constant that determines the average velocity of the trap ($v_{\text{avg}} = c/\Delta t$). The trap displacement is uncorrelated with the bead position.

For the information engine, the trap position is updated using the feedback algorithm

$$\lambda_{n+1} = \lambda_n + \alpha\,(x_n - \lambda_n)\,\Theta(x_n - \lambda_n)\,, \tag{8.2}$$

where $\Theta(\cdot)$ is the Heaviside function and α the feedback gain. In contrast with the power-stroke engine, the trap displacement is correlated with the bead position.

8.3 Energy Flow Measurements

The trap work (W, Eq. (2.33)), heat (Q, Eq. (2.34)), and potential energy (F, Eq. (2.35)) are measured using the equations discussed in Sect. 2.4.2. The integrated energy flow of the power-stroke engine as a function of time is shown in Fig. 8.2a. We find that the trap does work on the system to raise the bead. The slope of the trap work (blue) is higher than that of the stored gravitational energy (black). The difference in energy is dissipated as heat to the environment. The measured heat is shown in red.

The integrated energy curves, averaged over many trajectories, for the information engine are shown in Fig. 8.2b. The feedback gain was tuned to $\alpha = 1.7$ so that the trap does zero external work on the bead (blue). We find that the bead stores gravitational energy (black) even when trap work is zero, on average. The red curve denotes the heat dissipated by the bead to the environment. The negative slope of the curve means the heat is absorbed from the environment and converted into potential energy, while cooling the thermal bath.

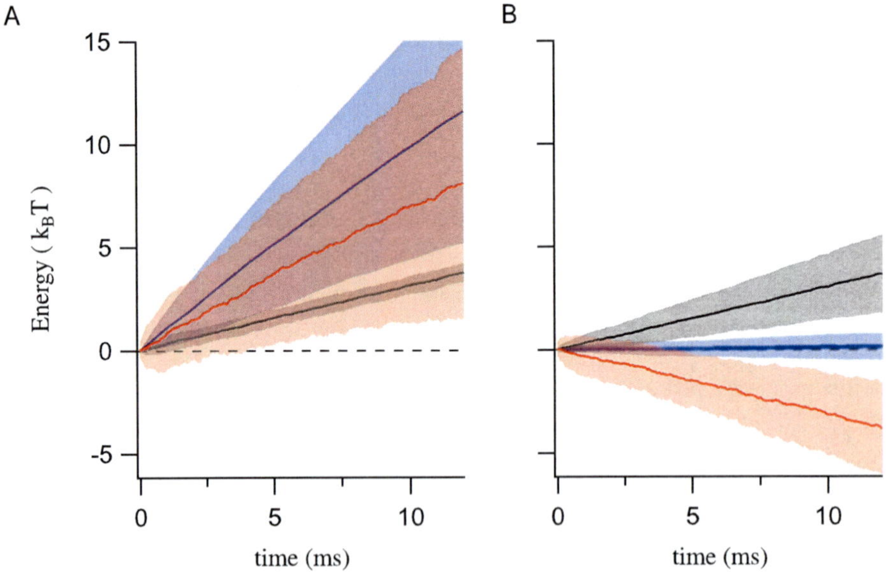

Fig. 8.2 Energy flows in (**a**) power-stroke and (**b**) information engines. The solid curves denote the experimental data of integrated work $\rangle W \rangle$ (blue), heat $\rangle Q \rangle$ (red), and stored gravitational energy $\rangle F \rangle$ (black). The shaded regions denote one standard deviation from the trajectory mean

8.4 Power-Stroke Engine Characterization

Figure 8.3 shows the integrated average trap work (W, blue), heat (Q, red) and stored gravitational energy (F, black). The shaded regions denote the standard error of the mean integrated energies. The first law of thermodynamics states the sum of all the energies flowing into and out of a system is conserved. The orange curve in Fig. 8.3 denotes the difference of the averages $\rangle W \rangle - (\rangle Q \rangle + \rangle F \rangle)$. The difference is zero, on average. The energy is thus conserved in the power-stroke engine, and all the fundamental energies have been accounted for. The additional operational costs, though large, can be neglected.

The energy flow in a power-stroke engine has been exactly calculated by Mazonka and Jarzynski [8]. Consider a bead, in gravity, translating in a harmonic potential at a constant velocity v. The probability density function of the bead, in the trap-centre coordinate, follows a Gaussian distribution $\mathcal{N}(\mu, \sigma^2)$, whose mean and variance are given by (cf. Eq. (2.52))

$$\mu = -\left(\frac{\Delta m\, g}{\kappa} + \frac{\gamma v}{\kappa}\right) \tag{8.3a}$$

$$\sigma^2 = \frac{k_\mathrm{B} T}{\kappa}, \tag{8.3b}$$

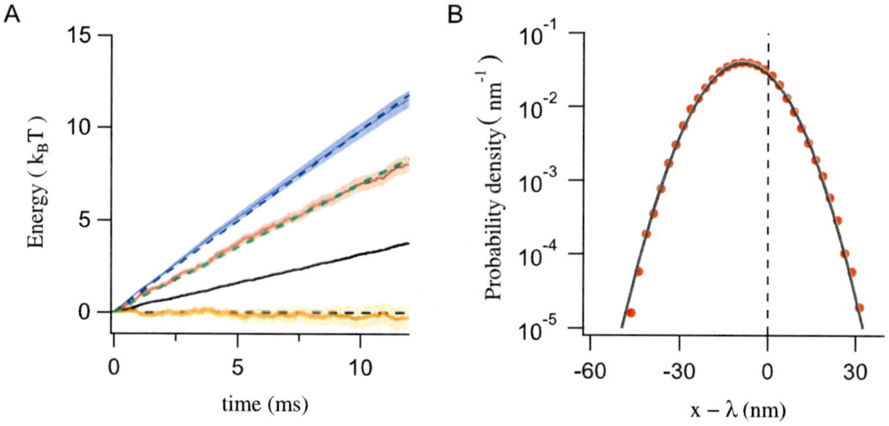

Fig. 8.3 Power-stroke engine. (**a**) Conservation of energy in a power-stroke engine. The solid curves denote experimental values of $\rangle W \rangle$ (blue), $\rangle Q \rangle$ (red), $\rangle F \rangle$ (black), and $\rangle W \rangle - (\rangle Q \rangle + \rangle F \rangle)$ (orange). The green dashed line is the plot of Stokes heat. The shaded regions denote the standard error of the mean. (**b**) The solid red markers denote the probability density of the experimental values of the bead displacement from the trap centre, $x - \lambda$. The grey curve is a fit to the Gaussian distribution $\mathcal{N}(\mu, \sigma^2)$, with a least-squares fit giving $\mu = -8.39 \pm 0.05$ and $\sigma = 10.06 \pm 0.04$. The experiments were performed at trap stiffness $\kappa = 50.1 \pm 0.5$ pN/μm, and velocity $v = 10.12 \pm 0.03$ μm/s

where κ is the trap stiffness and γ the coefficient of friction. The mean μ has contributions from gravity (first term) and velocity of the trap centre (second term).

The distribution of bead displacements from the trap centre $(x - \lambda)$ is shown in Fig. 8.3b. The experiments were performed using a 3 μm bead at a trap stiffness $\kappa = 50$ pN/μm. The velocity of the trap centre is $v = 10.1$ μm/s. The grey curve is the plot of a Gaussian distribution, whose mean and variance are given by Eq. (8.3). The mean gravitational potential energy of the bead, translating at velocity v, is $F = \Delta m\, g v t$, where t denotes time. The mean steady-state trap work is given by (cf. [8])

$$W = \kappa v \left(\frac{\Delta m\, g}{\kappa} + \frac{\gamma v}{\kappa} \right) t\,. \tag{8.4}$$

The corresponding heat dissipated by the bead is given by the Stokes drag

$$Q = \gamma v^2 t\,. \tag{8.5}$$

The green dashed line in Fig. 8.3a denotes the plot of Eq. (8.5). Similarly, the blue dashed line is the plot of Eq. (8.4).

The heat dissipated by the bead is negative for the information engine. Information engines can operate on a system without generating heat in the system. Information processing is fundamentally dissipative [9, 10]; however, the information processor can in principle be isolated from the work system. In addition, the information-processing protocol can be carried out at a lower temperature, which would lead to a lower information cost.

By contrast, power-stroke engines always dissipate energy to the thermal bath. In Appendix C, we show that heat is dissipated even when the trap follows the same trajectory that was recorded for the information engine (feedforward protocol), without any correlations with the bead position. Further, I perform experiments on an engine where the trap follows a drift-diffusion equation whose mean and variance are matched to those of the information engine (see Sect. 8.6). In both cases, heat is dissipated to the thermal bath.

Stokes drag, which causes heat dissipation, also creates hydrodynamic flows around the bead. Because the heat flows are different in the two cases, the hydrodynamic flows in general should differ, as well. However, as we show in Appendix B, any differences are too small to observe.

8.5 Probability Distribution of an Information Engine

In Fig. 8.4, we compare the probability distribution of the bead, in the trap coordinate, for the information engine to that of the power-stroke engine. The distribution of the information engine (blue) is skewed because the engine rectifies the fluctuations that cross the trap centre; the bead position is correlated with

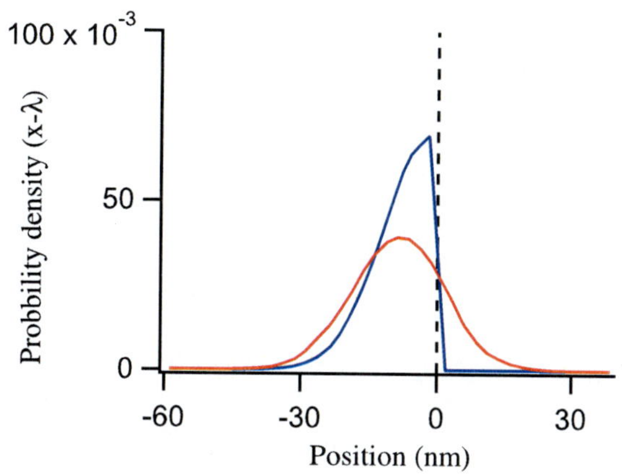

Fig. 8.4 Experimental data of the probability distribution of displacements from the trap centre an information engine (blue), compared to that of a power-stroke engine (red) translating a bead at a constant velocity, as in Fig. 8.2

the trap centre. Note that for the information engine, we plot $x_n - \lambda_{n+1}$ because of the experimental delay of one time step between a position measurement and the resulting trap-position update. In Appendix B, we show that the position distributions of feedforward and drift-diffusion engines are also uncorrelated. We conclude that knowing the bead and trap positions allows us to distinguish between an information and power-stroke engine without computing heat flows. There is no need to measure the heat flow directly, which would be difficult in this case.

8.6 Drift-Diffusion Engine

In this section, we study the energy flow and statistical distribution of an engine undergoing drift and diffusion. The trap centre is displaced according to the rule

$$\lambda_{n+1} = \lambda_n + c + \nu_n \,, \tag{8.6}$$

where ν_n is a Gaussian random variable that follows $\rangle \nu_n \rangle = 0$ and $\rangle \nu_m \nu_n \rangle = D_\mathrm{e}\, \delta_{mn}$, and D_e is the diffusion constant of the engine. The value of D_e is chosen to match the variance of the bead position powered by the information engine, calculated from the date presented in Fig. 8.2.

The energy flow of the drift-diffusion engine is presented in Fig. 8.5a. The work done by the trap (W, blue) is positive, and the bead dissipates heat (Q, red) to the environment. Note that the slope of heat is steeper than the power-stoke engine presented in Fig. 8.3 because of excess dissipation arising from additional motion of the bead in the medium. The distribution of the bead position with respect to the trap is shown as a black curve in Fig. 8.5b. The distribution of the drift-diffusion engine remains Gaussian distributed, in contrast to the information engine (blue).

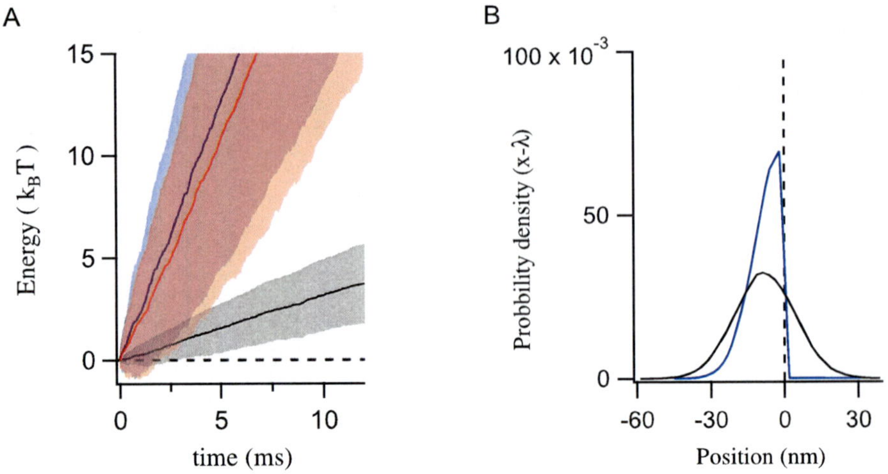

Fig. 8.5 Energy flow and statistics of an engine undergoing drift-diffusion. (**a**) Conservation of energy in a drift-diffusion engine. The solid curves denote experimental values of $\rangle W \rangle$ (blue), $\rangle Q \rangle$ (red), and $\rangle F \rangle$ (black). The shaded regions denote one standard deviation. (**b**) Probability distribution of an information engine (blue), compared to that of an engine undergoing drift-diffusion motion (black)

8.7 Displacement Distribution

In Sect. 8.5, we saw that it is straightforward to distinguish between an information engine and a power-stroke one when both the bead and trap positions may be observed. But what if only the bead position can be measured? This restriction seems likely when looking at biological motors, where the equivalent of a trap position might not be apparent. In this case, identifying an information engine becomes non-trivial. In particular, this constraint is of practical importance as it is usually challenging to measure the configuration/state of the molecular motor directly. Rather, one uses a probe whose position can be externally monitored. For example, the motion of a bead attached to a kinesin motor can be easily measured (compared to the position of the motor itself).

In Fig. 8.6, we present the skewness of the displacement distribution of the information (blue), power-stroke (red), and drift-diffusion (black) engines, measured at different time intervals. The displacement distribution of the power-stroke and drift-diffusion engines remain Gaussian (zero skewness) even at long time intervals ($\Delta t/\tau_r > 1$), where time is scaled by the relaxation time of the bead in the trap τ_r. However, for the information engine, the displacement distribution is skewed at long time intervals, showing a correlation between the bead's current position and that at an earlier time. The probability distributions of the bead's position displacements are presented in Fig. 8.7. At $\Delta t = 0.04\tau_r$ (displacements at one sampling time), the distributions are identical and Gaussian distributed. At $\Delta t = 10\tau_r$, the information

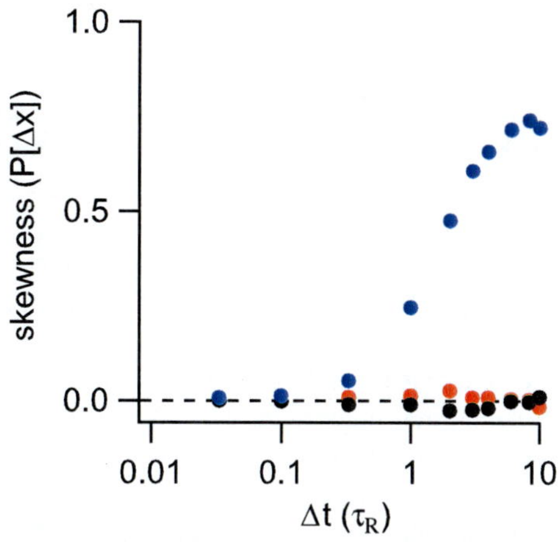

Fig. 8.6 Skewness of the probability density function of bead displacements ($\Delta x = x(t + \Delta t) - x(t)$) measured at different intervals Δt (in units of τ_r) for the power-stroke (red), drift-diffusion (black) and information (blue) engines

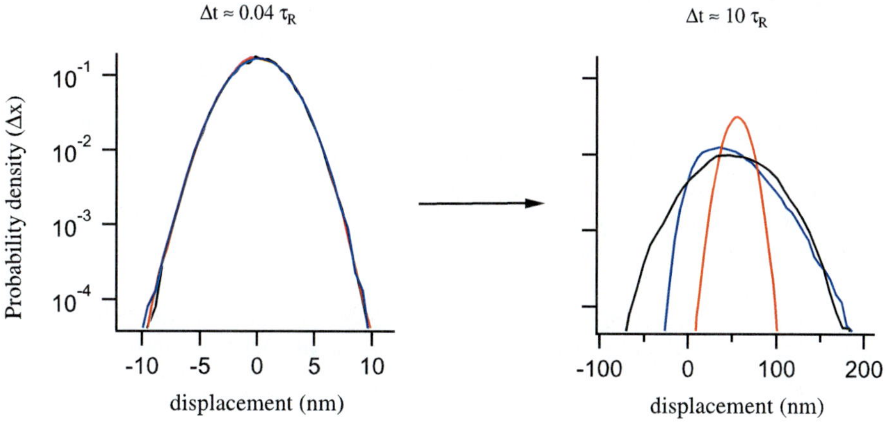

Fig. 8.7 Evolution of the displacement ($\Delta x = x(t + \Delta t) - x(t)$) distribution of the bead for the power-stroke (red), information (blue), and drift-diffusion (black) engines

engine (blue) distribution is skewed, whereas the power-stroke and drift-diffusion engines are Gaussian distributed. Note that the variance of the drift-diffusion engine is higher than that of the power-stroke engine because of the excess variance arising from the diffusive motion of the trap.

The skewness of the information engine is reminiscent of the bead's position distribution in the trap coordinate [Fig. 8.4]. At small Δt ($\ll \tau_r$), the bead's displacements arise mainly from Gaussian thermal noise and thus give rise to nearly Gaussian distributions with negligible skewness. By contrast, the long-time displacement ($\Delta t > \tau_r$) reflects the ratcheting, which truncates positive displacements once they cross the threshold. Note that probability density functions "bulging to the right" have positive skewness, while those bulging to the left have negative skewness.

8.8 Conclusion

I have shown that the information engine operates by converting thermal fluctuations in the medium into useful energy. In this chapter, I have contrasted some of the properties of an information engine against the simplest version of power-stroke and drift-diffusion "engines" that drag a bead without feedback. They use large amounts of energy to raise a bead as they simultaneously dissipate energy into the environment. I showed that if one has access to both the bead and trap positions as a function of time, an information engine can be easily identified by measuring the heat dissipated by an engine: An information engine cools the medium, whereas conventional engines heat it. When the trap stiffness or trap position is unknown, the information engine can be identified from the probability distributions that indicate correlations in the bead's motion. Here I distinguish the information engine from simple power-stroke and drift-diffusion engines. In general, the trajectory of a motor can be more complicated. Further work is needed to distinguish an information engine from other heat engines with more general motor dynamics. Also, the information engine presented here is based on a harmonic potential. The quadratic form of such a potential is special, as it means that the fluctuations of the displacement from equilibrium in a steady state produced by a constant force lead to a Gaussian probability density function, with zero skewness. Thus, the skewness of the displacement distribution produced by the information engine is contrasted against a "zero-skewness background" of the power-stroke and drift-diffusion engines. If the potential confining the particle were not symmetric, the distributions would always have skewness, and one would lose the qualitative distinction that exists in the simple case studied here. These studies can help identify motors that operate as information engines or as a combination of information and heat engines.

References

1. T.K. Saha, J.N.E. Lucero, J. Ehrich, D.A. Sivak, J. Bechhoefer, Maximizing power and velocity of an information engine. Proc. Natl. Acad. Sci. U.S.A **118**, e2023356118 (2021)
2. T.K. Saha, J. Bechhoefer, Trajectory control using an information engine. Opt. Trapp. Opt. Micromanipul. XVIII **11798**, 117980L (2021)
3. G. Paneru, D.Y. Lee, T. Tlusty, H.K. Pak, Lossless Brownian information engine. Phys. Rev. Lett. **120**, 020601 (2018)
4. B. Lau, O. Kedem, J. Schwabacher, D. Kwasnieski, E.A. Weiss, An introduction to ratchets in chemistry and biology. Mater. Horiz. **4**, 310–318 (2017)
5. W. Hwang, M. Karplus, Structural basis for power stroke vs. Brownian ratchet mechanisms of motor proteins. Proc. Natl. Acad. Sci. U.S.A **116**, 19777–19785 (2019)
6. D.J. Tritton, *Physical Fluid Dynamics* (Springer, Berlin, 2012)
7. K. Sekimoto, *Stochastic Energetics* (Springer, Berlin, 2010)
8. O. Mazonka, C. Jarzynski, Exactly solvable model illustrating far-from-equilibrium predictions (1999). cond-mat/9912121

9. R. Landauer, Irreversibility and heat generation in the computing process. IBM J. Res. Dev. **5**, 183–191 (1961)
10. C.H. Bennett, The thermodynamics of computation—a review. Int. J. Theor. Phys. **21**, 905–940 (1982)

Chapter 9
Conclusion

Information engines operate by converting heat in the surrounding medium into energy. They have refined our understanding of the second law of thermodynamics and our understanding of information as a physical quantity. Previous experimental studies were directed towards showing that these engines operate in a way that is consistent with the second law of thermodynamics, even though they may superficially seem to violate it. However, in this thesis, our goal has been to understand and optimize the performance of information engines. Information engines differ from molecular motors and heat engines: They use feedback to extract energy directly from the surrounding environment, paying only the cost of running the measurement and control apparatus. By contrast, molecular motors and heat engines use the energy from another source of fuel to operate.

We experimentally realized an information engine using an optically trapped bead. Our spring-mass model of the information engine simplifies the working principles of an information engine. The information engine converts the fluctuations of a thermal bath into stored gravitational potential energy. We showed that the work done by the spring on the mass can be set to zero by carefully raising the position of the spring. Then the gravitational energy is extracted purely from information about the thermal fluctuations that displace the mass. By contrast, raising a mass up in a viscous fluid, without using an information engine, requires input work by the spring. In steady state, the total input work equals the gain in gravitational potential energy and viscous dissipation.

We characterize the performance of the information engine, as quantified by either the rate of stored gravitational energy or velocity. We found that measuring and rectifying favourable fluctuations of the bead as quickly as possible maximizes the engine's performance. The rate of stored gravitational potential energy is further improved by increasing the values of material parameters such as the trap stiffness and bead size (and density). By optimizing the experimental parameters, we achieved output power and velocity comparable to those measured in molecular motors and marine bacteria, respectively.

T. K. Saha, *Information-Powered Engines*, Springer Theses,
https://doi.org/10.1007/978-3-031-49121-4_9

Information engines depend on the accuracy of measurements to apply appropriate feedback. We found that the output power decreases with the signal-to-noise ratio of measurements and has a phase transition to a zero-output regime below a critical signal-to-noise ratio. The phase transition is a novel (and unexpected) result that we found from our studies on information engines. By contrast, using a Kalman-filtered position estimate for making feedback decisions allows operation at all noise levels and can significantly improve the engine's performance near the critical signal-to-noise ratio. Furthermore, this approach also removes the phase transition in the information engine, a benefit of using Kalman filters that had not previously been recognized. We found that using a Kalman filter to apply feedback also improved the performance of information engines that either cool the thermal bath or transport a bead at an average velocity.

Although information engines can transport a bead at an average velocity, they do a poor job of controlling the bead trajectory because of stochastic perturbations. To solve this problem, we designed a new feedback algorithm that can control the bead position to follow a desired trajectory, while still satisfying the zero-work condition. We found that the performance of the information engine to track a desired trajectory is bounded by its material parameters, not by the feedback algorithm. This implies that better performance can be achieved by increasing the trap stiffness. A stiffer trap reduces the time between favourable fluctuations, for identical thermal noise strengths, which increases the output power and velocity of the information engine.

An alternative method of improving the performance of the information engine is to increase the strength of the fluctuations. Thus, we study the performance of the information engine in the presence of nonequilibrium noise in addition to equilibrium thermal noise. We find that the engine can output up to ten times more power than an information engine in a corresponding equilibrium thermal bath. The amount of increase was limited by the strength of the external fluctuations that we were able to create and can, in principle, be many orders of magnitude higher in realistic environments.

Finally, we considered the problem of distinguishing between information engines and other types of engines. Molecular motors are modelled as flashing ratchets, where the motor hydrolyses adinosine triphosphate (ATP) molecules to create conformational changes that lead to unidirectional motion. Previous experiments have demonstrated that information ratchets can perform better than flashing ratchets [1]. The similarity in operation of molecular motors to information ratchets raises the question whether molecular motors also use favourable fluctuations to improve their performance. Our results, which show an improved performance in an information engine's output in the presence of nonequilibrium noise, make it seem likely that such information engines may be found in cells, if one looks at the motor in the "right way". The cell has a strongly fluctuating internal environment. Why would its internal machinery not try to take advantage of such a resource?

Motivated by the goal of identifying molecular motors that sometimes or always operate as an information engine, we performed a preliminary study that distinguishes an information engine from a power-stroke engine. We find that the information engine extracts heat from the thermal bath and cools its surroundings.

By contrast, the power-stroke engine dissipates energy as heat into the environment. We also show that, when the dissipated heat cannot be measured, one can study the statistical properties of the bead position to distinguish information engines from power-stroke motor.

Although our method to distinguish information engines from other ratchet mechanisms by using the skewness of the displacement probability distribution works well for this simple case, the framework is too simple to apply directly to more complex systems. We still need to find a general framework to distinguish information engines from the other types outlined in Chap. 1 . Our study shows that it is possible to measure small correlations from highly stochastic trajectories. It was surprising that these correlations showed significant signals at long time separations.

9.1 Future Directions

In this section, we discuss the future directions for the research on information engines. We first discuss short-term scientific and technical-improvement projects. Then we discuss broader questions about information engines, in general, and about their impact.

9.1.1 *Short-Term Projects*

In this thesis, we built information engines that were designed to maximize their output. An immediate follow-up goal would be to quantify the information-processing cost of operating these engines. Our position detector measures "continuous-space" positions of the bead at discrete time intervals. Our apparatus simplifies the conceptual design of the information engine; however, measuring the corresponding information-processing costs directly in experiments is challenging. Still, models that account for physical constraints such as feedback delay could predict lower bounds on the information-processing cost. This can be used to infer upper bounds on the information-to-work efficiency.

Ian Ford proposed an experiment to measure the information-processing cost directly in continuous-position systems [2]. The engine consists of two spring-mass systems, with the masses coupled by another weak spring. The first spring-mass system operates as the work source (extracts energy), and the second system operates as the measuring device. Through its coupling with the work source, the measuring device estimates the state of the first system. The coupling strength controls the accuracy of the measurement and hence, the information-processing cost. Such a device can in principle be experimentally realized using the dual-trap apparatus introduced in this thesis. The beads in the dual trap can be modelled as two spring-mass systems, which are hydrodynamically coupled. The strength of the coupling could be controlled by changing the inter-trap distance.

A conceptually simpler next experiment would be to realize a Szilard-engine-like, integrated, two-state work source and measuring device system using a bead in a four-well potential. It can be generated by creating a virtual potential in two dimensions using an optical feedback trap [3, 4]. A double well along each axis would represent the work source and the measuring device. The potential could be manipulated to create correlations between the sub-systems. Then during the measurement step the state-measuring device indicates the state of the system. Once the states of the system and measuring device are coupled, the potential could be manipulated to extract the energy from the work system by exploiting the information obtained from the measurement. This setup and choice of potential can experimentally reach the expected bound on both information-processing cost and output energy, using a single system. I did some preliminary experiments and simulations (with Matthias Gey, not presented here) that demonstrate partially the concept.

The energy extraction is compensated by information processing in a Szilard engine. However, the energy extracted by an information engine can in principle be greater than the information-processing cost if the erasure/measurement is carried out at a lower temperature T_I, compared to the work source temperature T_S, as noted in Chap. 7. This can be experimentally verified by implementing the measuring device and the work source using the four-well potential. Then, for $T_S > T_I$, in the quasistatic limit one can extract energy equal to $k_B T_S \ln 2$ ($> k_B T_I \ln 2$), which is greater than the information-processing cost. The time-dependent temperature can be experimentally implemented with the current apparatus by using the electronic noise source. Further experiments that measure quantities such as the thermodynamic cost of a bit flip (a NOT operation), bounds on energy extraction and dissipation in a noisy Szilard engine can be performed using this apparatus.

Technical Improvements

Although the current apparatus is already quite flexible, several straightforward improvements can make the apparatus more robust for experiments with longer protocol time. First, the FPGA can be directly connected to the voltage amplifiers, to apply electro-kinetic feedback forces on the bead [3, 5], in addition to the optical feedback forces. This technique can be used to study bead dynamics in complex total potentials, using uncoupled optical and electrokinetic forces, and to measure energy changes arising from each source, individually. Experiments on a Szilard engine can be performed without changing the trap stiffness and bead diameter between measurements, which would otherwise require force and position re-calibrations. Thus, the experiments can be programmed to automate force and position calibration and run all the protocols. Long protocols require experimental stability against mechanical drifts over the protocol time scales. Mechanical drifts usually arise from faulty mirrors, after prolonged use of the mirrors to align the counter-propagating laser beams. The need to realign mirrors could be reduced by introducing the detection and trapping lasers into the sample chamber from the same side, using a single objective. This modification would simplify the alignment procedure because both lasers, when properly collimated, would focus in the same plane. Minor alignments, if necessary, could be made using mirrors on motorized stages. As both beams pass through the same optical components, they would drift together with no relative motion, thus reducing the frequency of re-alignments.

9.1.2 Longer-Term Goals

In this thesis, we presented a design of an information engine that is equivalent to a Szilard engine that extracts fluctuations in one direction only. The engine ratchets when there are upward fluctuations and does not move during downward fluctuations. However, Szilard imagined an engine that could extract fluctuations from either direction, storing them all as gravitational energy. An equivalent model of information engine needs to be designed for our system. Previous information engines have been designed to harvest energy from thermal fluctuations. This is equivalent to macroscopic energy-harvesting engines that are designed to convert wind energy into electricity. Our information engine is designed to maximize output energy rates, under different real-world constraints. Other energy harvesting engines may benefit from the design principles and feedback algorithms described in this thesis. Similarly, the design principles known to improve performance in macroscopic engines can be used to design better information engines. With miniaturization of electronics, one can imagine autonomous information engines that can be designed to move in a medium, using thermal fluctuations.

Our studies have shown that information engines operating in nonequilibrium environments can extract large amounts of energy, and their performance improves with the strength of the fluctuations. As molecular motors always operate in nonequilibrium environments, it is possible that they also utilize the surrounding fluctuations to enhance their performance. It will be interesting to identify molecular motors that operate partially or completely as information engines.

References

1. B.J. Lopez, N.J. Kuwada, E.M. Craig, B.R. Long, H. Linke, Realization of a feedback controlled flashing ratchet. Phys. Rev. Lett. **101**, 220601 (2008)
2. I.J. Ford, Maxwell's demon and the management of ignorance in stochastic thermodynamics. Contemp. Phys. **57**, 309–330 (2016)
3. M. Gavrilov, Y. Jun, J. Bechhoefer, Real-time calibration of a feedback trap. Rev. Sci. Instrum. **85**, 095102 (2014)
4. A. Kumar, J. Bechhoefer, Nanoscale virtual potentials using optical tweezers. Appl. Phys. Lett. **113**, 183702 (2018)
5. A.E. Cohen, W.E. Moerner, Method for trapping and manipulating nanoscale objects in solution. Appl. Phys. Lett. **86**, 093109 (2005)

Appendix A
Other Information Engine Models

Here we discuss models of information engines that convert thermal energy to other useful quantities. We show that the Bayesian information engine framework improves the performance of these information engines, as well.

A.1 Transporter Information Engine

The information engine can also be used to transport a bead using thermal fluctuations, as presented in Chap. 5. In Chap. 6, we saw that measurement noise reduced the rate of free-energy extraction for a bead that is lifted against gravity. If the noisy measurements are used naively, there is a phase transition as a function of signal-to-noise ratio (SNR). Below a critical value $\mathrm{SNR_c}$, it is not possible to have a "pure" information engine where no work is done directly on the system. By contrast, using a Bayesian information engine, we can extract work for all values of SNR. For $\mathrm{SNR} \approx \mathrm{SNR_c}$, the Bayesian engine harnesses a significant fraction (order 50%) of the energy that can be extracted when there is negligible measurement noise.

In this section, we show that the situation is similar when one considers directed horizontal velocity (normal to gravity). In Fig. A.1, we show that the Bayesian information engine can increase the bead's velocity. In this experiment, the bead is trapped horizontally and transported perpendicular to the direction of gravitational force.

The experiments were performed using a 3 μm bead at trap stiffness 38 pN/μm and diffusion constant $D = 0.17\,\mu\mathrm{m}^2/\mathrm{s}$. The scaled mass $\delta_{\mathrm{g}} = 0$, as the bead is transported perpendicular to gravity. The velocities of the naive (red) and Bayesian (blue) transporter engines are shown in Fig. A.1a. We find that the velocity generated by the Bayesian information engine is higher than that of the naive engine at

T. K. Saha, *Information-Powered Engines*, Springer Theses,
https://doi.org/10.1007/978-3-031-49121-4

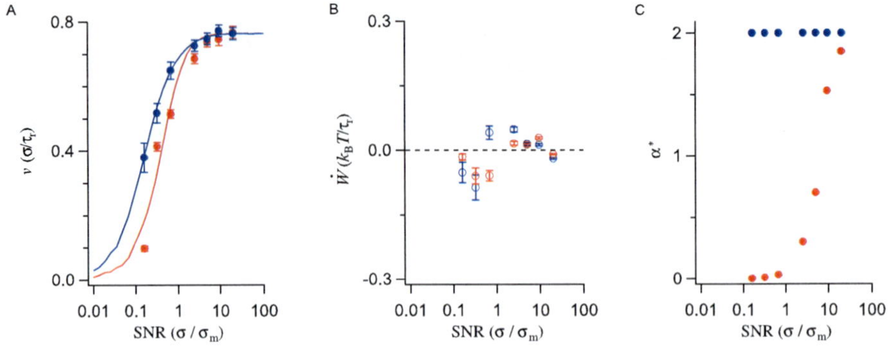

Fig. A.1 Performance of naive and Bayesian transporter engines. (**a**) Transport velocity, (**b**) input trap work and (**c**) feedback gain α for the naive (red) and Bayesian (blue) transport engines. The markers denote experimental values, and the solid lines are numerical simulations performed by Joseph N. E. Lucero [1]. The error bars denote the standard error of the mean

intermediate SNR. The corresponding work done by the trap is shown in Fig. A.1b. Figure A.1c shows the feedback gain used to set $\dot{W} = 0$.

A.2 Cooling Information Engine

In this section, we compare the performance of the information engine when operated as a refrigerator to "cool" the bath. The engine consists of an optically trapped 3 μm bead, where the trap is shifted orthogonal to gravity such that $\delta_g = 0$. The experiments were performed for trap stiffness $\kappa = 40$ pN/μm and diffusion constant $D = 0.17\,\mu\text{m}^2/\text{s}$. The position of the trap is updated using one of two feedback rules (naive or Bayesian),

$$\lambda_{n+1} = z_n \,, \tag{A.1}$$

where $z_n = \{y_n, \hat{x}_n\}$, y is the noisy measurement, and $\hat{x}$ is the Bayesian estimate. The trap is re-centred to the measured y or estimated bead position $\hat{x}_n$ when the engine is powered by the naive or Bayesian engine, respectively. The information engine rectifies thermal fluctuations in the bath to raise the bead's position in the trap potential. Resetting the trap using Eq. (A.1) cools the bath by transferring heat from the bath to potential energy of the bead in the trap potential. The extracted energy is calculated using

$$F_{n+1} = -\frac{1}{2}\left[(y_{n+1} - \lambda_{n+1})^2 - (y_{n+1} - \lambda_n)^2\right]. \tag{A.2}$$

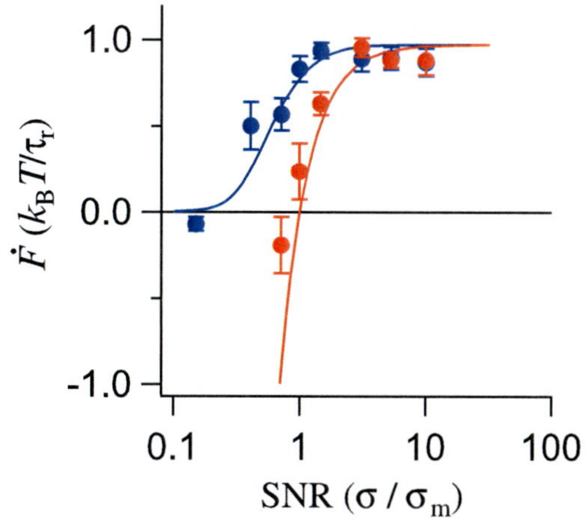

Fig. A.2 Performance of naive and Bayesian cooling engines. The rate of heat extracted from the thermal bath for the naive (red) and Bayesian (blue) cooling engines. The markers denote the experimental value and the solid lines are the theoretical predictions calculated by Jannik Ehrich (personal communication). The error bars denote the standard error of the mean

Note that the extracted free energy is the negative of the trap work defined in Eq. (6.4). The heat extracted by the Bayesian (blue) and naive (red) information engines is presented in Fig. A.2. For SNR<1, the naive information engine dissipates energy into the thermal bath because of feedback errors. By contrast, the Bayesian engine can still extract energy from the thermal bath and work as a refrigerator. At low SNR, the Bayesian engine's performance also deteriorates and saturates to $\dot{F} \to 0$. The situation is similar to that shown in Fig. 6.6.

Appendix B
Hydrodynamic Flows

When a bead is dragged through a viscous fluid, it generates hydrodynamic flow fields in the medium. What happens to those flows when the bead is ratcheted and the heat is extracted from the bath, instead of being dissipated into the bath, to generate directed motion? Here we show experimentally that the hydrodynamic flow fields in the two situations are experimentally indistinguishable.

To measure the hydrodynamic flows around the bead, we built a dual-trap apparatus. We trap a single 3-μm bead in each of the two traps, whose centres are separated by 7 μm. The positions of the two beads in the dual trap are hydrodynamically coupled [2]. The position of the first (driving) bead is moved perpendicular to gravity along a trajectory that follows $A \sin \omega t$, where $A = 82.5$ nm, $\omega = 2\pi f$ and $f = 10$ Hz. The positions of both the driving and probe beads are recorded using a video microscope, with a frame rate of 500 Hz. The cross-correlation function of the probe bead arising from hydrodynamic interactions created by the driven bead is derived in Ref [3]. The amplitude of the cross-correlation response, $C = |\chi_{21}|A$, at the driving frequency ω_0, is [3]

$$C = \frac{\omega_0 \mu' k_d A}{\sqrt{(\text{Det}M - \omega_0^2)^2 + \omega_0^2 (\text{Tr}M)^2}}, \tag{B.1}$$

where ω is the frequency of the driving bead, κ_1 is the trap stiffness of the first (driving) trap, A is the driving amplitude and Det$M$ and TrM are the determinant and trace of the matrix M, respectively. The experiment was repeated for trap separations of 13 and 20 μm. The experimental data is presented in Fig. B.1 as red markers and the solid curve is the plot of Eq. (B.1).

Next, the driving bead follows the trajectory $A \sin \omega t$, for the same amplitude $A = 82.5$ nm and frequency $f = 10$ Hz, but now powered by an information engine. The input trap work is zero. The feedback algorithm follows Ref. [4]. We repeat the cross-correlation analysis based on video microscopy. The measured amplitudes of the cross-correlation for the different trap separations are shown as blue markers in

T. K. Saha, *Information-Powered Engines*, Springer Theses,
https://doi.org/10.1007/978-3-031-49121-4

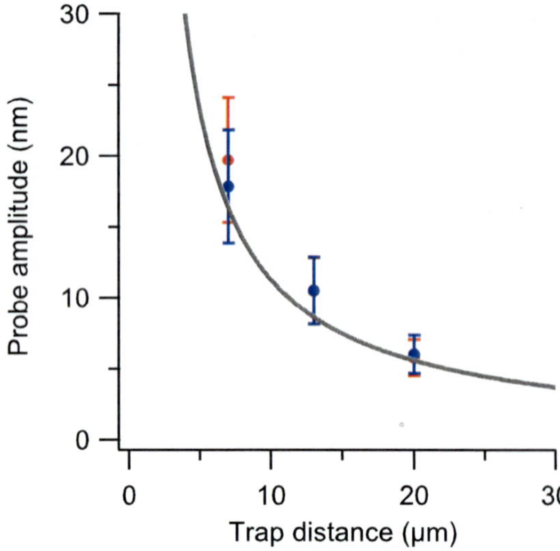

Fig. B.1 Experimental cross-correlation amplitude of the probe bead for power-stroke (red) and information engine (blue) driving of the driving bead for different centre-to-centre distance between the static trap centres. The error bars denote the standard error of the mean. The grey curve is a plot of Eq. 7 for $\kappa_d = 36.0$ pN/μm and $\kappa_p = 0.8$ pN/μm

Fig. B.1. To within our experimental resolution, we find no difference between the hydrodynamic flows generated by the two types of engine.

The heat dissipated by the bead, for the power-stroke engine, rapidly diffuses away, leading to a negligible temperature difference.[1] That is, we do not expect that temperature differences between the two situations are large enough to alter the hydrodynamic flows appreciably. Although the bead moves through the medium without dissipation, the bead exchanges energy and momentum with the fluid. The information engine modifies the force field such that the trap does no work that is dissipated as heat. The trap position is modified such that it biases the forward motion of the bead. Thus, for the same average velocity, the hydrodynamic flows produced by the power-stroke and information engines are indistinguishable.

[1] The temperature difference ΔT create by a 3 μm bead dissipating heat at a rate $\dot{Q} = 10^{-19}$ W, in water, can be estimated by a model of a sphere of radius r producing heat at a rate $\dot{Q}$ in a medium of thermal conductivity k, which leads to $\Delta T \sim \frac{\dot{Q}}{rk} = \frac{10^{-19}}{1.5\times10{-}6\times0.5} \approx 10^{-14}$ K.

Appendix C
Feedforward Engine

In this section, we present the energy flow in a feedforward engine where the position of the trap is updated based on a previously recorded trap trajectory of an information ratchet. In this case, the trap dynamics is identical to that of an information engine; however, unlike the case of an information ratchet, the bead position is independent of the trap position. We measure the energy flow in the feedforward engine and compare it to the information engine, as shown in Fig. C.1. We find that heat is dissipated by the feedforward engine, to raise the bead. Thus, negative heat dissipation is a robust signature of an information engine: The difference in heat flow arises *solely* from the correlations between the motion and the thermal fluctuations of the surrounding bath.

T. K. Saha, *Information-Powered Engines*, Springer Theses,
https://doi.org/10.1007/978-3-031-49121-4

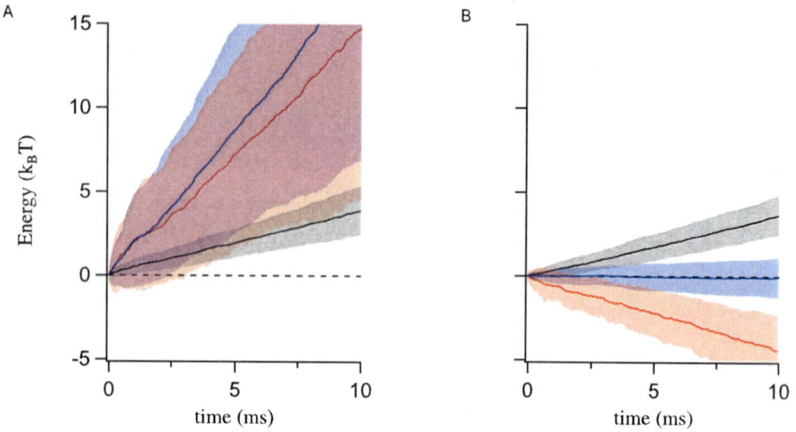

Fig. C.1 Energy flows in (**a**) feedforward, and (**b**) information engines. The solid curves denote the experimental data of integrated work $\langle W \rangle$ (blue), heat $\langle Q \rangle$ (red), and stored gravitational energy $\langle F \rangle$ (black). The shaded regions denote one standard deviation from the trajectory mean

References

1. T.K. Saha, J.N.E. Lucero, J. Ehrich, D.A. Sivak, J. Bechhoefer, Bayesian information engine that optimally exploits noisy measurements. Phys. Rev. Lett. **129**, 130601 (2022)
2. J.C. Meiners, S.R. Quake, Direct measurement of hydrodynamic cross correlations between two particles in an external potential. Phys. Rev. Lett. **82**, 2211 (1999)
3. S. Paul, R. Kumar, A. Banerjee, Two-point active microrheology in a viscous medium exploiting a motional resonance excited in dual-trap optical tweezers. Phys. Rev. E **97**, 042606 (2018)
4. T.K. Saha, J. Bechhoefer, Trajectory control using an information engine. Opt. Trapp. Opt. Micromanipul. XVIII **11798**, 117980L (2021)